IN OCEANS DEEP

ALSO BY BILL STREEVER

*And Soon I Heard a Roaring Wind: A Natural History
of Moving Air*

Heat: Adventures in the World's Fiery Places

Cold: Adventures in the World's Frozen Places

Green Seduction: Money, Business, and the Environment

Saving Louisiana? The Battle for Coastal Wetlands

IN OCEANS DEEP

Courage, Innovation, and Adventure

Beneath the Waves

BILL STREEVER

Little, Brown and Company

New York Boston London

Little, Brown and Company
Hachette Book Group
1290 Avenue of the Americas, New York, NY 10104
littlebrown.com

First Edition: July 2019

Little, Brown and Company is a division of Hachette Book Group, Inc. The Little, Brown name and logo are trademarks of Hachette Book Group, Inc.

Unless otherwise noted, all images are within the public domain.

The publisher is not responsible for websites (or their content) that are not owned by the publisher.

The Hachette Speakers Bureau provides a wide range of authors for speaking events. To find out more, go to hachettespeakersbureau.com or call (866) 376-6591.

ISBN 978-0-316-55131-1
LCCN 2018965114

10 9 8 7 6 5 4 3 2 1

LSC-C

Printed in the United States of America

*For my late father, who always supported my
love of the world beneath the waves.*

O abyss! O eternal Godhead! O deep sea! What more could you have given me than the gift of your very self?

—Saint Catherine of Siena

So my message is, in whichever realm, be it going into space or going into the deep sea, you have to balance the yin and yang of caution and boldness, risk aversion and risk taking, fear and fearlessness.

—James Cameron, "Titanic and Other
Reflections," 2004

Every person on this earth depends on the sea in some way. My gosh, if only we could get folks interested in that part of our planet, the benefits could be unlimited. But first, someone has to notice.

—Bob Barth, U.S. Navy Sealab diver, 2000

CONTENTS

DISCLAIMER

This book is not a diving manual. Safe diving in all of its many forms requires training provided by internationally recognized organizations. Nothing in this volume should be construed as encouraging anyone to ignore widely accepted standards for education, equipment, depth limitations, experience, and other factors that minimize risks.

IN OCEANS DEEP

PREFACE

My father, who was my first diving partner well before I was old enough to drive, built his working life around running midsize businesses. But when I was barely out of high school and found my first job diving, he was more than supportive. He was enthusiastic. On several occasions, I overheard him bragging to friends and colleagues about his son the oil field diver, the kid making his living underwater with explosives and cutting torches, the young man assembling pipelines on the seabed.

Ten years later, when I walked away from a successful run as a diver to pursue a university degree, he was disappointed. This was a legitimate reaction. My work had taken me from the Gulf of Mexico to the South China Sea. It had provided an opportunity to dive while breathing exotic gases and to live at depth with three colleagues for monthlong stretches. My future looked far less exciting. And to this day, it seems at times that one of my biggest mistakes was turning my back on a career underwater.

On the other hand, I had suffered several diving injuries and survived a few close encounters with my own mortality. I had grown tired of the time I was spending at sea, as much as eight months during busy years. During quieter periods, the uncertainty of living from contract to contract and weathering lean seasons without a paycheck took its toll. And there were two supervisors I respected who quietly encouraged me to get out of the business while I was still young, to pursue an education, something that had eluded both of them despite their abundant intelligence and talent.

And so I went to college. I became a biologist and a writer. I ran research programs in Australia and Alaska. I continued to lead a reasonably interesting life. But it could never compete with what had come before. So whenever possible, I inserted myself into field studies that required diving or the use of underwater sampling gear, things such as dredges, grab samplers, nets, and even hydrophones that recorded the vocalizations of passing whales and the noises from ships. And I dived recreationally, usually with scuba gear, whenever and wherever I could. Diving remained, and will continue to be, an important part of my day-to-day existence.

Back to my father. When my book *Cold* was reviewed in his favorite magazine, the *Economist,* he urged me to write something about diving. Later, as Parkinson's disease ravaged first his body and then his mind, he grew adamant.

I told him there were problems with such a book. Many great works had already covered the topic, but few had found a wide readership. And there was the question of whether my style of writing suited the subject. Most of what I had read on underwater exploration tended toward either personal exploits or something close to a textbook, but as an author I was partial to narratives shaped by science and history.

After several years of reflection, I realized that what I had to write could not be limited to diving in the conventional sense but instead would tackle what I came to think of as "humanity's presence beneath the waves." When friends asked about my work, I would repeat those exact words. And then I would explain myself, adding something like, "You know, diving, submarines, submersibles, and underwater robots."

Even then, I had to consider how to capture the essence of my subject while also captivating readers. I had to figure out how to do justice to this amazing story. Well into the first year of research and interviews for what became *In Oceans Deep*, my approach remained undetermined, until one day, while I was free diving, something close to inspiration struck. Afterward, of course, the book continued to evolve, and very late in the process it went through a major change following an interview with the world's most famous female diver and underwater conservationist, Dr. Sylvia Earle.

I never intended to write an unabridged treatise on humanity's presence underwater. I meant instead to create the kind of book that would leave a lasting impression. I wanted readers to finish the last page with a newfound or rejuvenated motivation to free dive, to scuba dive, to consider passage in any one of the increasingly common tourist submersibles capable of reaching beyond one thousand feet, to watch undersea documentaries, to consider buying an underwater robot of their own, and to talk to marine scientists and submariners and the engineers whose work lies beneath the surface of the sea. In other words, I wanted readers to embrace the part of our world that is shrouded by depth.

I continue to hope that *In Oceans Deep* will bring attention to one of the most important yet under-discussed topics of our time. Past accomplishments and the feats being attempted today are barely known to the general public. Even among those who work underwater, knowledge is far too compartmentalized. For example, champion free divers may know nothing about submersibles, and submersible pilots might know very little about the strange gas mixtures that allow divers to breathe below one thousand feet. And the link between the ways in which humans access the underwater world and the struggle to protect our troubled oceans deserves far more consideration than it receives.

While my father passed away before I even began to write this book, I was able to finish it only because of him. The reality that he is not here to appreciate the result is among the tragedies of my life. But for those who see value in these pages, please know that at least half the credit goes to my late father. Without his occasionally nagging encouragement, *In Oceans Deep* never would have come to be.

Chapter 1

DESCENT

Early on January 23, 1960, a thirty-seven-year-old Italian engineer named Giuseppe Buono looked out at the Pacific and worried.

After two days of towing the ungainly submersible *Trieste* onto location above the Challenger Deep in the Mariana Trench, the waves were growing. Here, 220 miles from the western Pacific island of Guam, floating far above the deepest point in the world's oceans, wave heights exceeded five feet. While seas of that size are mere ripples to an oceangoing surface ship, *Trieste* was built for calmer waters. On the surface, she was a fragile vessel.

Buono had prepared *Trieste* for her sixty-four previous dives, and on this day he did not like what he saw. *Trieste*'s deck was awash. The shallow-water telephone, used by *Trieste*'s pilots to communicate with her support crew at the beginning of each dive, had been swept away. The sea had also destroyed an instrument that measured speed of descent and ascent. And *Trieste*'s vertical current meter hung from its mount, dangling from wires, swinging through the air as the vessel bobbed up and down.

From his vantage point aboard the support ship, Buono did not know what else might be broken or lost. The waves and the damage they had already caused did not bode well for a record-setting dive to the ocean's deepest floor, seven vertical miles down and—if all went well and catastrophe did not prevail—seven vertical miles back up.

According to written accounts describing the day's events, on that morning in 1960 Buono was "taut with anxiety." But Buono was not calling the shots. U.S. Navy lieutenant Don Walsh was the Navy's officer in charge, a formal title meaning that he was in command. Civilian Jacques Piccard, who had worked with *Trieste* long before the Navy acquired her, offered important advice. Buono was the engineer, but Walsh and Piccard were the two men who would be aboard *Trieste* for her historic journey. They were the test pilots, and they were not the sort of men who succumbed to anxiety.

The submersible Trieste *before her 1960 dive to the deepest point in the ocean. The people in the small launch behind* Trieste *provide a sense of scale. Most of what is visible below the waterline is* Trieste's *gasoline-filled buoyancy chamber.* (Vallejo Naval and Historical Museum)

If *Trieste* was to dive, the crew had to act quickly. The descent would require around five hours, and the ascent would need an additional three hours. The plan also called for thirty minutes on the seabed itself. Walsh and Piccard, though hardly risk averse, did not relish the prospect of surfacing after dark in heavy seas. And they did not want anyone to have to attach the one-inch-thick towing cable to *Trieste* after dusk, a tricky job that required a swimmer in the water.

Trieste suspended by a crane. The personnel capsule—the pressure-proof sphere that carried Don Walsh and Jacques Piccard to and from the bottom of the Challenger Deep—is attached to the bottom of the striped buoyancy chamber. (NH Series, NH 9600, NH 96801, Archives Branch, Naval History and Heritage Command, Washington, DC)

"I am going to check the main electric circuits in the sphere," Piccard told Buono. "Then, if everything is in order, we shall dive immediately."

What he called the sphere was the personnel capsule attached to the bottom of *Trieste*. It was a hollow ball of steel that would protect him and Walsh from the almost unthinkable pressures found below.

～

In 2016, I meet Don Walsh—the deepest man alive—in person, at his home in rural Oregon. He lives with his wife on a property well inland from the coast. Not counting a son who resides in a separate house on the same property, his nearest neighbor is a mile away.

Don is eighty-four years old, but his upright posture, the surety of his gaze, his pleasure at the sight of a black bear climbing an apple tree at the edge of his long driveway, and his energy of engagement with me—a stranger—all suggest someone far younger.

I am very pleased to meet him. It means something to me to stand next to him in the flesh, while time allows.

Fifty-six years earlier, the year before I was born, Don Walsh and Jacques Piccard descended to the deepest known point in the world's oceans, seven miles below the surface. No one would repeat this feat until 2012, when filmmaker and explorer James Cameron would follow in their footsteps, taking his submersible *Deepsea Challenger* to a depth just slightly shy of that reached by *Trieste*. To this day, only three human beings and two robots have visited the Challenger Deep. In contrast, twelve people have been to the moon and well over five hundred have traveled in space.

For perspective, if Mount Everest were somehow magically relocated to the deepest known point in the world's oceans, the summit would be submerged, its peak standing some six thousand feet beneath the surface.

For further perspective, 1960 was one year after the YMCA began offering scuba diving instruction in the United States, four years before the Navy's Sealab experiments kept divers at depth for days at a time using a technique that would become known as saturation diving, a decade before rapidly changing economies and technologies would push the world's search for oil and gas far out of sight of land, twenty years before tethered robots would revolutionize underwater work and exploration, and fifty-five years before untethered robots relying on artificial intelligence would begin to proliferate in the world's oceans.

Standing in the entryway to his house, I fill Don in on my own background underwater and on my current life living aboard a cruising sailboat with my marine biologist wife, sailing from place to place to better understand, firsthand, the oceans. I also tell him of my desire to write a book that will resonate with anyone who has ever looked out from a beach or over the side of a boat, and wondered, *What's under those waves?*

I am too old to admit to having heroes, but standing there, still in his entryway, I tell Don Walsh that he is and has been, for as long as I can remember, one of my heroes.

"Aw, shucks," he replies, timed and toned so as to sound not only anachronistic but also self-effacing and amused. Then he leads me through his home and up a flight of stairs to his office, a wonderfully large, open space designed by his wife, Joan, lined in lightly stained bookshelves holding eight thousand volumes—books on naval history, philosophy, submarines, Bill Bryson's *A Walk in the Woods*. And there we sit talking for four hours without a break.

Trieste, named after the Italian town where she was built, was not exactly a submarine. She was, in some ways, more akin to the kind of sealed diving bell used by the likes of William Beebe to reach a depth of 3,028 feet on August 15, 1934, twenty-six years before the Challenger Deep expedition. But diving bells are winched up and down from the depths as they hang ignominiously from the end of a long wire tether, rendering them unwieldy and subject to currents. Sealed diving bells protect their occupants from the pressure, but they are incapable of reaching the bottom of the Challenger Deep.

The pressure-proof sphere just big enough for two men that hung from *Trieste*'s underside had some things in common with a diving bell, but *Trieste* suffered neither the indignity nor the limitations imposed by a cable. She explored the depths unencumbered by surface connections.

Trieste's inventor, the remarkable scientist Auguste Piccard—who also happened to be Jacques Piccard's father—was known for his use of balloons in high-altitude research, and the U.S. Navy described *Trieste* as "the underwater equivalent of a lighter-than-air craft, much like a blimp operating in reverse." She was a diving bell attached to an oblong balloon designed to fly through the sea.

In a true dirigible, hydrogen or its less explosive cousin helium, both lighter than air, provides the buoyancy that allows flight. The pilot can gain altitude by adding gas to buoyancy chambers. The gas displaces air, and the dirigible floats upward. To return to earth, the pilot dumps gas from the chambers.

For *Trieste*, meant to fly at the deepest of depths, gas was not an option for buoyancy control. The abyss comes with great pressures. Every 33 feet of depth in seawater adds 14.7 pounds per square inch of pressure, the equivalent of the pressure generated by the earth's atmosphere at sea level. At 66 feet beneath the surface of the ocean, the pressure is equivalent to that generated by three of the earth's atmospheres at sea level, or about 44 pounds per square inch. At 297 feet down, the pressure increases to ten times that found at sea level, or about 147 pounds per square inch. *Trieste* was headed far deeper. At the bottom of the Challenger Deep, the equivalent of 1,100 of the earth's atmospheres would press against *Trieste*'s sphere with a force of about 16,883 pounds per square inch.

To understand what this might mean, consider that a typical scuba tank is filled with ordinary air at 3,000 pounds per square inch. Under rare circumstances, a scuba tank can explode, its compressed contents suddenly ripping through its metal walls, releasing about the same amount of energy that would come from the simultaneous detonation of two hand grenades. An exploding scuba tank, suddenly releasing its contents, would destroy everything around it. But a full scuba tank dropped into the Challenger Deep would not explode. Instead, it would be crushed long before it reached the bottom.

In ordinary submarines, pressurized air is used to control buoyancy. Submariners can flood ballast tanks to submerge and then force air into those same tanks to surface. But available high-pressure gas bottles, the kind that hold the air used for buoyancy control in submarines, would be of no use at the bottom of the Challenger Deep. Even if they were not destroyed by the crushing weight of extreme depths, when opened they would simply fill with water, providing no buoyancy whatsoever.

Trieste adhered to the principles of dirigibles, but she filled her balloon—what Piccard and others called her "float"—with liquid gasoline. Unlike gases, liquids are not easily compressed. In response to the very high pressures of the depths, along with the colder temperatures, gasoline would lose only seventeen percent of its volume. And it is lighter than water. It floats. Even under the tremendous pressures found on *Trieste*'s itinerary, gasoline would retain most of its volume and therefore most of its buoyancy.

Trieste's float was fifty feet long and twelve feet wide, not much bigger than a typical sailing yacht. It was divided into twelve compartments, two that could be filled with compressed air for use on or near the surface and ten for the gasoline.

But how did she descend? She carried sixteen tons of iron shot as ballast. Initially, near the surface, she released air from two compartments in her float. She also could release a portion of the gasoline from her float. Crew members—including the engineer Giuseppe Buono—sometimes stood on her deck to add extra weight for those first few feet departing from the surface. Ultimately, the weight of her ballast, her sixteen tons of iron shot, pulled her downward. Later, when the time came to ascend, she dropped iron shot.

Why iron shot? Why not lead? Because iron shot could be controlled magnetically. If things went wrong, electromagnets would shut down and the iron shot would fall away. *Trieste,* free of her ballast, would ascend. Even if the power failed, whether the two men inside were conscious or unconscious, alive or dead, *Trieste* would carry them back to the surface.

She was a strange contraption, a hybrid, neither a submarine nor a diving bell, nothing more than a gasoline-filled float carrying a sphere with a diameter of less than seven feet. If all went as planned, the sphere would protect Don Walsh and Jacques Piccard from the mind-boggling pressures that are the very essence of the Challenger Deep. Its walls were five inches thick. Two acrylic viewports, cone-shaped and six inches thick, were all that allowed *Trieste*'s occupants to look outside, to see the wonders of the depths. But the high pressures necessitated the smallest of ports; they were a mere four inches across inside the sphere. The view for *Trieste*'s test pilots would be something like the view through a narrow tunnel.

Trieste's inventor and those who used her called her a bathyscaphe—"bathy" for deep and "scaphe" for hull or for vessel, boat, or ship. While she looked nothing like a surface ship, this undersea dirigible—carrying about 34,200 gallons of gasoline, sixteen tons of iron shot ballast, and her tiny sphere just big enough to hold two very cramped men—was indeed a ship of the depths.

And while not exactly a death trap, neither was she what anyone would call inherently safe. She would not come close to meeting modern safety standards. But this was a time when individuals had more say regarding their own risk tolerance.

Fifty-six years later, in Don's office, I ask about their contingency plan. What would they have done, all those years ago, if *Trieste* had, for example, become stuck in the mud seven miles beneath the waves, held in place even after dropping her iron ballast? Don smiles. Becoming stuck in the mud, he recalls, was a real possibility. It would have been one of many ways to die in *Trieste*. He and his surface support crew joked about a bouquet of flowers kept in a freezer for just such an occasion. The crew had strict instructions to spread the flowers on the sea if *Trieste* failed to surface.

Everything about *Trieste* exhibited the sort of innovative risk-tolerant thinking required to touch down on the bottom of the Challenger Deep. It was the same sort of thinking that had led to the development of diving helmets, scuba gear, and submarines; of what came to be known as saturation diving and atmospheric diving suits; of submersibles—including some homemade models—that could not dive as deep as *Trieste* but that were far more mobile. And it was a move away from this risk tolerance, but not from innovation, that led to a world in which undersea robots, controlled from the surface or working autonomously, now play an ever-greater role in exploration of the depths.

As I sit talking to Don Walsh, thinking of what I know and guessing about what I have yet to learn, I think about the story at hand. If only Walsh and Piccard had waited their turn, if only humanity had taken one small step at a time into the abyss, building one depth record on top of another, the story would be simple. I could start at the top and work down, if only that were how it had happened.

As it was, *Trieste* jumped ahead. But there is more to the story of the deep sea than putting two men on the bottom within the confines of a pressure-proof sphere.

I am not the only person in recent years to interview Don Walsh about *Trieste,* but I am one of only a few. Another was Chris Wright, who talked to Don for *No More Worlds to Conquer,* a book that explores the lives of people who have done something extraordinary and then faded from the limelight.

Wright asked Don about his life after *Trieste.* "Well," Don reportedly answered, "a lot of people think I died."

But of course he did not die. He served in the Vietnam War, commanded a military submarine, earned three graduate degrees, worked as deputy director of Navy Laboratories, dived on the *Titanic* and the *Bismarck,* started and ran the Institute for Marine and Coastal Studies at the University of Southern California, and founded a consulting company. Now—and bear in mind that he is eighty-four years old—he has just returned from promoting personal submersibles to a client base of billionaire megayacht owners in Monaco. Not long before that, he was in Switzerland helping Rolex with a documentary. And before that, he was at sea for several weeks.

When we talk, his answers tend to wander comfortably through memories, multifaceted but entirely coherent, reflecting the complexity of the life he has led and continues to lead.

I have one burning question for Don Walsh: What happened? Not with the dive itself, which is well documented, or with his own life, which is described in various articles and in Chris

Wright's book. Instead, what happened with the exploration of the deep sea? After the *Trieste* dive to the bottom of the Challenger Deep, no one went back until 2012. And the 2012 expedition was itself an isolated event, not part of a national quest. It was as if Neil Armstrong took one giant leap for mankind and then the United States turned its back on the moon.

"One thing," he says, "is NASA. The space program overshadowed the government's undersea program. NASA had flames and good photography and cool space suits. What we did with *Trieste* was very hard to convey to people. It's dark down there. There are no stunning vistas. Just two guys huddled in a tiny sphere."

He is not—adamantly not—critical of NASA and its many successes. "I don't want to give the impression that I am against the national space program," he says. "I am only asking that the inner space program on planet Earth get some parity with outer space. The ocean community has gotten a lot from the space program in the way of technology transfer. They can do research and development work at a scale we cannot afford. And due to NASA's interest in the several other oceans in our solar system, some much larger than our world ocean, they have been supporting a considerable amount of Earth-ocean science on Earth." But that is not even close to parity.

"There is no question that the United States and other major maritime nations need to dramatically increase support and coordination to explore our planet's largest geographic feature," Don says.

He explains that the *Trieste* dive almost never happened. "We were just fourteen people working on the program. I think the average age was twenty-two or twenty-three years old. I was twenty-seven. Piccard was in his thirties." He talks, too, of the importance of others on the team, men like Dr. Andreas Rechnitzer and Lieutenant Larry Shumaker, both critical to the success of *Trieste* but often left out of the historical accounts. This is not a story of Walsh and Piccard alone, but rather one of a team effort, a tale of fourteen men at sea and many others ashore. And all of them relied on unsung predecessors. There were, before *Trieste,* other bathyscaphes, also invented by Auguste Piccard but funded by the Belgians and later the French. Don knew then, as he knows now, that he was one person among many.

The official U.S. Navy had little interest in what they were doing. Senior officers did not want to be held responsible for what seemed to border on foolhardiness. This was not war. *Trieste* and her dives were all about providing scientists with access to the realm they studied, part of what some people call the "research and development Navy," at least one step removed from any immediate military purpose.

At the level of the Navy's chief of operations, the project had been approved, but that is not to say that it had the universal support of mid-level management.

"We operated in a well-funded vacuum," Don recalls. And he liked it that way. It gave him and the rest of the team freedom to innovate.

He was told, among other things, not to publicize the mission. If they succeeded—equivalent to "if you survive"—there might be room for publicity, but the Navy did not relish the possibility of a publicized failure.

Just before the dive, with *Trieste* awash in the waves and beaten up by two days of towing through a rough chop, someone in the chain of command contacted the support vessel. Don was aboard *Trieste,* so the mission's chief scientist, Dr. Andreas "Andy" Rechnitzer, took the message. The program supporting *Trieste,* the message said, had been canceled. It was time to pack up and sail for port.

In relation to this communication, Don asks if I am familiar with Horatio Nelson's blind eye. The British admiral lost the use of one eye when enemy gunfire hit a nearby sandbag, sending debris into his face. Later, during the Battle of Copenhagen in 1801, Nelson was attacking the enemy. The admiral commanding the British forces was aboard a nearby ship. Using flags, the admiral signaled for a retreat. According to one biography, Nelson turned to the man standing next to him in the heat of the battle and said, "I have only one eye, and I have a right to be blind sometimes." He held his telescope to his blind eye and pointed it in the general direction of the signal flag before pressing on to victory.

With *Trieste* bobbing in the waves, going through final checks just before its historic dive, Rechnitzer wandered around the support vessel. He stepped into the wardroom for a cup of coffee, considering his reply. While he deliberated, *Trieste* slipped beneath the surface.

"Unable to comply" read Rechnitzer's response to the message saying the program had been canceled. "*Trieste* is passing 10,000 feet." They were not going to abandon the program now, even if doing so required a little white lie, a small exaggeration, a Nelson's blind eye.

⌒

Back in 1960, just before Rechnitzer received the message that he would artfully ignore, Piccard climbed through the entrance tube that stretched from *Trieste*'s upper deck, on top of her gasoline-filled float, down to the sphere, the personnel capsule. He checked the bathyscaphe's main circuits, including those controlling the electromagnets that held the iron ballast.

He climbed back out. Despite the concerns of *Trieste*'s engineer, Piccard was ready to dive. Walsh gave his approval, and by extension that of the Navy.

Normally, the two men in the sphere would count on the shallow-water telephone to guide their initial descent. They would close their hatch, call for the flooding of the entrance tube, and request to be released from the surface support vessel. If anything went wrong near the surface, as the descent began, the telephone would issue a call to abort the dive. But the telephone was gone, destroyed during the tow through rough seas.

"When I have closed the door," Piccard told Buono, referring to the hatch of the sphere, "you may open the entrance-tube valves and proceed with normal operations. If, at the last moment, something doesn't go well, I shall turn the propeller, and you will know that we must give up the dive."

The propeller that Piccard would use to signal a problem, the propeller on this ship of the deep, was mounted on deck, on top of the gasoline-filled float. Buono would be able to see it easily from the surface support ship until *Trieste* was completely submerged.

The bathyscaphe rolled with the waves. The motion was not quite like that of an ordinary ship. It was the kind of roll to be expected of a gasoline-filled float, of what Piccard called "a play thing of the waves." For those accustomed to the sea, it was not uncomfortable. There was no question of seasickness. But nevertheless, the two men were ready to submerge, to drop below the perpetually moving surface.

As things worked out, they did not have to turn the propeller. There was no need to signal Buono, to give up the dive. They were on their way to the bottom of the sea.

The descent began at 8:23 in the morning. It went slowly. "Ten minutes after leaving the surface," Piccard later wrote, "we were at a depth of only 300 feet."

There the bathyscaphe stopped. It had reached a thermocline, the top of a layer of suddenly colder water. Just as water has a greater density than air, cold water has a greater density than warm water. *Trieste* was floating on cold water.

Walsh and Piccard could feel *Trieste* moving up and down, rising and falling. The motion was not caused by the surge from the seas far above their heads, which would not be noticeable at this depth, but from the motion of the thermocline itself, from what Piccard called "internal waves," rollers moving below the ocean surface.

At three hundred feet, they were still shallow, and yet they had already passed the practical limit of humans breathing compressed air—of ordinary divers breathing air from scuba tanks or from long umbilicals attached to compressors on the surface.

The two men had to make a decision. They could wait for the gasoline in their float to match the temperature of the surrounding water. As its temperature dropped, its density would increase and the descent would recommence. But this would not happen quickly. Not wanting to risk the possibility of exceeding their planned dive duration, of surfacing later than expected, they decided to release some of *Trieste*'s gasoline. Like balloonists venting hot air from a balloon, they dumped gasoline into the sea. *Trieste* resumed her slow descent.

After dropping a mere thirty-five additional feet, *Trieste* halted again. She had found another thermocline. The men liberated more gasoline.

Five minutes passed, and ninety feet of additional depth were gained before *Trieste* halted again, on top of another cold-water layer. Still more gasoline went into the sea, and *Trieste* descended to 530 feet, where she found yet another thermocline, causing one more unexpected pause.

"This was the first time in my 65 dives in the *Trieste*," wrote Piccard, "that I had observed this phenomenon of repeated stratification."

From the perspective of Walsh and Piccard, and of *Trieste*

herself, 530 feet was not especially deep. In fact, it was shallow. But it approximates the boundary of survival for humans breathing compressed air—not the practical boundary, but the extreme daredevil limit, the usually fatal beyond-stupidity depth. While more than one person has reached 500 feet while breathing air and lived to surface, the dives are so dangerous that even the good people of Guinness World Records want nothing to do with them. In 2005, the record keepers stopped recognizing depth records by divers breathing compressed air.

To return to *Trieste* and her famous dive: more gasoline went into the water. The needle on *Trieste*'s depth gauge recommenced its downward creep.

Trieste passed 831 feet, the current record for a human being diving on a single breath of air, set by Austrian free diver Herbert Nitsch in 2012. He was assisted in the descent by a heavy weight and on the ascent by a balloon. This is another form of daredevil diving, extremely dangerous, not even remotely possible for most people, often deadly even for those with the natural abilities and training required for such an attempt. On his record-setting dive, Nitsch lost consciousness while riding his balloon upward. Although he awoke before reaching the surface, he soon complained of severe weakness, and he suffered paralysis. Despite extensive treatment, it was initially thought that he would be paralyzed for life. Since then, he has gradually recovered, and he continues to dive today, although he no longer pursues records.

Trieste passed 1,090 feet, the current record for a scuba diver breathing exotic gas mixes that negate most of the physiological problems of pressurized air.

At 1,500 feet, Piccard and Walsh put on warmer clothes. Below the thermoclines, the tropical sea was not especially tropical. The outside temperature would soon drop to forty degrees Fahrenheit. Steel can hold back the pressure, but the water's chill oozes through.

Two hundred and fifty-two feet later, at 1,752 feet, *Trieste* passed the deepest depth reached by human divers breathing compressed gas in the ocean. Those divers did not use scuba tanks. They wore helmets and breathed a strange mixture of helium, hydrogen, and oxygen, a gas mix unknown to nature on this planet or, for that matter, any other planet. But even with their boutique gas mix and their elaborate support system, they faced challenges. With gas at this pressure, simply inhaling and exhaling was laborious. It was as though they were breathing pea soup. In addition, they had to deal with something called helium tremors, which made their limbs move uncontrollably. And their joints did not work as they would on the surface. On top of all of that, when they finished on the bottom, they had to decompress for more than thirty days, slowly letting helium and hydrogen dissolved in their blood escape through their exhalations as they gradually moved into shallower water. During that time, the divers sat in a pressurized steel cylinder called a decompression chamber. They read, they slept, and they breathed, off-gassing.

Trieste's descent continued.

At 2,000 feet—not yet one-tenth of the way to her destination—the sea outside was dark. Few of the sun's photons penetrate below 2,000 feet even in the clearest of seas.

At 2,200 feet, Walsh and Piccard, looking through their tiny viewports, saw flashes of luminescence. Many of the creatures at this depth, mostly gelatinous things, are equipped in the manner of lightning bugs and fox fire fungus. They glow in the dark.

Trieste passed 2,400 feet. Although the U.S. military will not confirm precise depth capabilities for modern American military submarines, 2,400 feet is probably well beyond their crush depth, the point at which their pressure hulls fail and all hands are lost. Yet for *Trieste*, 2,400 feet was still shallow. *Trieste* was not made for such shoal-water work.

Walsh and Piccard watched their depth and their clock. Since

23

the instrument that normally measured the rate of descent had been taken by a wave, they had to calculate it on their own.

Occasionally, they released iron shot, slowing *Trieste*'s free fall.

At 3,000 feet, they passed the depth at which air-breathing leatherback sea turtles sometimes swim in their relentless hunt for soft-bodied prey, for jellyfish and salps and ctenophores.

Nine hundred feet deeper, at 3,900 feet, they passed the deepest recorded dive of a sperm whale, the gigantic creature pursued by Herman Melville's obsessed Captain Ahab.

Walsh and Piccard passed 4,500 feet. In 1960, no one was fishing in 4,500 feet of water, but in the coming decades, as shallow-water fish stocks were depleted, commercial fishing fleets moved deeper. In the right ocean at the right time, a submersible at this depth could encounter a deep sea trawl harvesting 150-year-old fish, plowing through colonies of strange deepwater corals, leaving in its wake a seabed bulldozed to death.

At 5,000 feet, Walsh and Piccard passed the depth of what would become the world's first approved deep sea mine, which would be proposed for a place called Solwara 1, near Papua New Guinea. As planned, robots would extract ore rich in gold, silver, and copper.

About one hundred feet deeper, they passed the bottom depth of the BP oil well that would eventually spill an estimated 219 million gallons of oil into the Gulf of Mexico. That well, at that depth, would not be alone. The only thing that would set it apart from others of its kind was that it would spectacularly and publicly fail, becoming a botched job that would or should remind everyone paying attention that hubris and the deep sea, even decades after *Trieste,* remain incompatible.

An hour and fifteen minutes into the dive, at 5,600 feet, more than a mile down, the crew of *Trieste* placed what Piccard referred to as a "telephone call." It did not rely on the shallow-water telephone, which of course had been lost during towing. Instead, it went through what Piccard termed a "sonic telephone" and what Walsh described to me as a voice-modulated sonar. Radio waves cannot penetrate far in seawater, but communication is possible using sound waves. The call was a routine matter, as much to check the functioning of the telephone as to communicate about their progress.

As the descent continued, Walsh and Piccard heard the whir of their fan blowing in the sphere, pulling air through a canister of soda lime, a chemical that absorbed their exhaled carbon dioxide. Occasionally, they listened to a quiet hiss as one of them opened a valve to inject oxygen into their little atmosphere, replacing that lost to carbon dioxide. There was also the sound of their own voices calling out numbers and the scratch of their pencils as they made notes and ran hurried calculations. And there were the grunts, groans, and squeaks of steel ever so slightly changing its shape as it adjusted to the increasing pressure.

Walsh and Piccard placed two more telephone calls from *Trieste*, at 10,000 and 13,000 feet, both routine. Really, there was not much to say. They were descending as planned. Aside from the concerns about surface weather, this dive was as predictably mundane as a record-setting dive could be.

They were gaining depth now at a rate of three feet per second.

At 20,000 feet, below three miles, they passed the depth limit of the tiny fleet of research submarines that would eventually replace *Trieste*.

At 24,000 feet, Walsh and Piccard broke their own previous record. "We are at a depth where no one has yet been," Walsh told Piccard in the twilight world of their tiny sphere.

They kept going.

Approaching the bottom at eleven thirty in the morning, they dropped more of their iron shot, slowing their descent in hopes of avoiding a sudden crash onto the seafloor.

When they turned on their exterior mercury vapor lamps, they saw nothing but clear water. To save battery power, they turned off the lights.

They waited, sinking slowly through absolute darkness.

At 32,500 feet, still far above the bottom, they heard a sudden cracking sound, and *Trieste* abruptly shuddered.

The two men looked at each other. For a moment they thought they might have hit bottom, but they realized that their depth sounder had detected nothing and they were still descending.

They turned off all of the bathyscaphe's equipment and listened, hearing a continuous quiet clicking that might, they thought, be shrimp, or possibly *Trieste*'s paint cracking under the pressure. But the descent continued smoothly.

"In my opinion," Piccard said, "it isn't anything serious; we are not losing any gasoline. Let's go on, and we'll see later."

Walsh responded with one word: "Okay." Later, he wrote that he was not worried about the possibility of hull failure. If *Trieste*'s sphere had been compromised, they would already have been dead.

At 12:56, almost five hours after they left the surface, *Trieste*'s depth sounder picked up a signal from the seabed. They had three hundred feet to go. Once again, they turned on the exterior lights. Outside, they saw thousands of tiny jellyfish and what Walsh remembers today as small shrimp.

They dropped more ballast, further slowing their descent, awaiting touchdown. It came ten minutes later. Meeting the bottom, *Trieste* stirred up a cloud of creamy silt and halted, all alone at what is known as full ocean depth.

"The *Trieste* took possession of the abyss," Piccard wrote, "the last extreme on our earth that remained to be conquered."

These words, in retrospect, were poorly chosen. *Trieste*, a visitor, a guest, a transient present on the seabed for a precious few fleeting moments, conquered nothing. Humanity was and to this day remains a stranger to the abyss.

The eighty-four-year-old Don Walsh chases his dog out of his office, sending it scuttling down the stairs.

I ask him why the government never pursued something akin to NASA for the deep sea. "You know," he says, "before I left the Navy, I became a deputy director at the Navy Laboratories. So I spent a lot of time thinking about money. I was a bean counter. That informed my outlook. That made me look at all the hands stretched out asking for money." An undersea NASA just never made the cut.

"If you had the opportunity to pitch a deep sea exploration program to the president," I ask, "what would you say? How would the one-minute elevator speech go?"

"By the 1970s," he says, "there were seven submersibles that could go to 20,000 feet or so. The Chinese own and operate the deepest submersible today, good for 23,000 feet. Twenty thousand feet gives you access to ninety-eight percent of the world ocean's seafloor."

Don pauses, considering the earth. "Two percent doesn't sound like much. But that two percent of ocean bottom that remains more or less inaccessible is about the size of the United States, including Alaska, plus half of Mexico. It includes all the great ocean trenches. But the burden of manned systems working at those depths is too high."

There is the cost factor, but there is also the risk factor. "Someone would wind up writing letters to widows," he says. "No one likes to write letters to widows."

Instead, in his view, the future will be in robotics. "All the great trenches need to be studied," he goes on, "but all the heavy lifting will be done by AUVs—autonomous underwater vehicles."

In 1960, small computers that would have been capable of assisting the two men who traveled in *Trieste* did not yet exist.

Walsh and Piccard might have benefited from something as simple as a pocket calculator to work out the mathematics of buoyancy and pressure and temperature and the rates of ascent and descent, but instead they used a pencil. Now, computers capable of controlling complicated undersea missions not only exist but are in routine use well removed from the nearest human. AUVs are tetherless robots, the computerized unmanned submersibles that have only just begun to proliferate, the latest chapter in humanity's story beneath the waves. Artificial intelligence, the stuff of science fiction during the *Trieste* dive, now swims in the depths.

"But manned submersibles will not always be off the table," he continues. "Even today, there are things that people can do better than machines. The idea of having one of us there is very important."

His elevator speech seems to be moving in a different direction. "What kid," he asks, "wants to grow up to be a robot?" But he stops there, the question rhetorical, the answer too obvious to state.

Just as *Trieste* touched bottom, Piccard peered through a viewport, one of the four-inch-diameter windows to the outside, and took advantage of the bathyscaphe's lights.

"Nature would have it that the *Trieste* came down on the bottom a few feet from a fish," Piccard later wrote, "a true fish, joined in its unknown world by this monster of steel and gasoline and a powerful beam of light. Our fish was the instantaneous reply (after years of work!) to a question that thousands of oceanographers had been asking themselves for decades."

The two men watched what would be described as a sole of some sort, a flounder, about twelve inches long and six inches wide. The apparition, Piccard later wrote, "moved away from us swimming half in the bottom ooze, and disappeared into the black night, the eternal night which was its domain."

He waxed poetic, but what he saw may not have been a flounder or even a fish at all. The deepest confirmed depth for a fish hovered two miles above his head, at 26,722 feet, and that specimen looked nothing like a flounder. It was a snail fish, an animal that resembles a deathly pale tadpole.

No one knows for sure what Walsh and Piccard really saw. They may have seen a deep-dwelling sea cucumber. Sea cucumbers, or holothurians, are relatives of starfish and sea urchins. The kind that live in shallow water look something like giant slugs. In certain regions, such as Japan, they are considered delicacies. Walsh and Piccard, with their vast experience in the ocean, would certainly have recognized ordinary sea cucumbers. But they might not have recognized the sea cucumbers sometimes found at extreme depths, the kind with flattened bodies and fins of sorts, the ones capable of swimming by undulation, skimming just above the bottom or even flying through the water. They might have looked at one of these creatures through the cloud of silt stirred up by their bathyscaphe and glowing beneath their mercury vapor lamps and seen exactly what they hoped to see, a fish, even though there were no fish living anyplace close by. But the possibility that it could have been a fish should not be dismissed. So little is known of life at these depths that the observation reported by the crew of Trieste may, in the end, turn out to be correct.

They also reported seeing something that looked like a reddish shrimp. Animals that look like shrimp do indeed occur at great depths. Walsh and Piccard may or may not have been mistaken about the fish, but they nevertheless observed life at the deepest of depths.

The plan called for Trieste to stay on the bottom for thirty minutes, taking observations. There, at the deepest of all seabeds, in the dimly lit sphere sandwiched between ivory-colored silt and a float full of gasoline, as they stared out through the viewport into the halo of *Trieste*'s external light, what went through their minds? They must have wondered about the ascent. Were they stuck in the mud forever? Was the crack and shudder they had heard and felt at 32,500 feet something that might kill them after all? Could something have happened to the float? When they cut the current to the electromagnets, would the ballast fall away as planned?

When it came to their survival, they knew the sea to be indifferent, impartial, entirely uncaring, capricious with respect to human intentions. If something went terribly wrong, the telephone, if it worked at this depth, would allow them to communicate, but it could not summon help. They occupied the only vessel capable of even approaching the bottom of the Challenger Deep. They were alone.

Decades later, as part of an Explorers Club interview, Walsh was asked about fear in the abyss. Was he ever scared "down there," the interviewer wanted to know.

"No," replied Walsh, "you're on your game, you're very alert. Being scared and having fear zaps your mental acuity and you can't afford to do that because you have to stay very sharp. All these practice dives we had been making in Guam were exactly the same, so by the time you reach the deepest dive, you become, I don't want to say 'one with your machine,' but close to it."

When asked about the dive in relation to courage, Walsh has also described it as "just a longer day in the office."

In written accounts, neither Walsh nor Piccard admitted to fear or anxiety on the bottom. It was as if their future survival was not immediately relevant. One with the machine or not, right then, at that moment, they were as alone as two people can possibly be, but they were also as alive as few people would ever be. They were alive in a place where no humans had been before. If they were dead tomorrow, so be it. At least they were there today, at that instant, in that place.

They measured the sea temperature at about thirty-eight degrees Fahrenheit and watched for signs of moving water, a measurable current, but saw nothing. They checked for indications of radioactivity but found none.

To test the telephone at this deepest of depths, Walsh called the surface. "Trieste on the bottom," Walsh recalls saying, "six thousand fathoms."

The surface responded, asking him to repeat his message. He did so. And he told them that *Trieste* would emerge from the depths around five o'clock in the afternoon, about an hour before the tropical darkness would close in on the ocean waves.

After the phone call, Piccard activated *Trieste*'s rear spotlight. Walsh looked through the viewport that faced *Trieste*'s entry tube, the same one they had crawled through five hours earlier to begin their dive.

"I know what happened," Walsh said to Piccard, "that noise, that jolt." He was referring to the excitement at 32,500 feet. "It was the big viewing port of the entry tube that cracked."

The entry tube had been flooded before *Trieste* left the surface, so its big viewing port had not succumbed to water pressure but rather to temperature changes. *Trieste*'s metal hull contracted at a different rate than the port.

The fracture posed no risk at depth, but it might be a problem on the surface. There, they would use compressed air to blow water out of the entry tube prior to exiting. The crack might have to be fixed by divers before they could blow the water out and leave the sphere hanging under *Trieste*'s float full of gasoline.

The men shivered in the deep sea's chill. They ate chocolate.

Although scheduled to rest on the seabed for thirty minutes, at twenty minutes they were ready to go. "We cast a final look upon this horizonless land," Piccard later wrote, "shining under the glow of our searchlights."

Despite Piccard's language, Walsh recalls that there was little to see outside. The cloud of creamy silt that had risen up from the seabed when *Trieste* touched down continued to reflect their lights. Visibility at times was in the neighborhood of two inches.

They released shot from one of the ballast silos, and the weights hit the soft sediment of the seabed and sent another billowing cloud upward and outward, a cloud within a cloud, visible only because it further dimmed the glow of the outside lights. But *Trieste* was once again under way, headed for the surface seven miles above their heads.

As she left the bottom with its pressure of 16,883 pounds per square inch, *Trieste*'s gasoline expanded ever so slightly. Her buoyancy increased. At first, she traveled at one and a half feet each second. At a depth of six miles, she was moving at two and half feet per second. At two miles, each second took her four feet closer to the surface, the afternoon daylight, the unrestricted air.

She accelerated to five feet per second, or three hundred feet per minute. Hours later, not far from the surface, she passed through the same thermoclines that had slowed her descent. But she continued upward, performing just as her crew and the crew of her support vessel expected.

"Being in a ship," wrote the great English author Samuel Johnson in the days of sail, "is being in a jail with the chance of being drowned."

Writing as he was in 1759, Johnson could have been familiar with submarines. The first true one was built in 1620 and tested in the Thames eighty-nine years before Johnson was born, but Johnson made no mention of the technology. Had he written something about those early submarines, he might have compared them to surface ships. "Being in a submarine," he might have quipped, "is being in the punishment block of the worst prison, its bad air reeking with the stink of humanity at close quarters, and not so much a chance of drowning as a chance of not drowning." Had he been knowledgeable about bathyscaphes, he might have compared them to punishment cells buried in the deepest holes of punishment blocks, tiny rooms of physical restraint, darkness, cold, and what to him, in his time, would have seemed to be the near certainty of drowning.

If the sea surface was as bad as a jail, the deep sea was a place of nightmares. The great depths, it was once supposed, could be nothing but sterile deserts, expansive plains devoid of life and possessing few if any interesting features. But the brutality of the depths is only true from the perspective of those accustomed to the surface. To creatures accustomed to extreme pressures, the shallows are hostile. Walsh and Piccard saw only hints of life, but parts of the deep sea, it turns out, support abundant life. Parts of it teem with living communities.

Some of that life would be of interest to scientists. Some of it would spawn a hypothesis claiming that life on earth began not in the shallows, as is commonly believed, but far below. Some of it would prove to be valuable to medical science. Submersible expeditions would be launched for no other reason than to seek out species whose own unique biochemistry produced substances that could be used to fight human diseases.

In contrast, many who lived on the surface would come to see the life at great depths as a tremendous nuisance. To those who viewed the oceans as a vast resource basin, a new economic frontier covering two-thirds of the planet's surface, virtually untouched and unexplored, deep sea life would prove to be a stumbling block that required environmental assessments, an irritant that led to criticisms from special interest groups, a rallying call for those who believed that government regulations should protect even parts of the globe that no one would ever see. The life at great depths would spawn an international regulatory body, an arm of the world government that would restrict certain activities.

All of this was yet to come. All of this was long after *Trieste*. And all of it is part of humanity's presence in the deep sea.

The eighty-four-year-old Don Walsh comments on the fish that he and Piccard thought they saw. "It was not a scientific dive," he says. "Neither of us were trained scientists. We were like test pilots. Our job was to make sure it was safe for scientists. We saw the fish just before we landed. It looked like a flatfish—a halibut, a sole, a flounder. Of course, we were pooh-poohed by everybody. They said it was a sea cucumber. But in the past year, fish have been seen at great depths. So maybe time is on our side. I won't say we'll be vindicated, because I never went to the wall on this, but I think people will see that there are fish at great depths."

And then he continues his elevator speech, but now backing away from a direct human presence on the seabed and returning to robots, to AUVs. "There are vast reaches of the ocean that remain unexplored," he says. "We send people to the moon, but eighty-five percent of the world ocean is unknown. That is awful. The classic way to do ocean exploration is surface ships, but there is not enough money in the world to do the rest of the oceans with surface ships. To me the clear answer is AUVs, which I call dumb research ships. They can do things without humans."

He envisions a contest for the first autonomous underwater vehicle to circumnavigate the globe. He has proposed calling it the Captain Cook Challenge. AUVs have already crossed the

Atlantic and the Pacific, so the next leap forward, as he sees it, might be an AUV that can go all the way, nonstop, one way, from Point A to Point A.

Eventually, for less than the cost of a single research vessel, a future fleet of deep-diving AUVs could spend days and weeks at a time cruising the ocean basins, sending back data, chipping away at the unknowns.

⌐⌐

Three hours and seventeen minutes after leaving the bottom, *Trieste* bobbed on the surface, once again wallowing in the waves. Thinking of the cracked viewport, Walsh and Piccard bled air slowly into the entry tube. For fifteen minutes, they watched as the water level slowly dropped. Their worries came to naught. Without mishap, they opened the hatch separating their sphere from the tube. They climbed out, returning to the surface world.

Learning of the successful dive, the Navy's mid-level management changed course. Suddenly, they became enthusiastic boosters of *Trieste* and her crew. "The purpose of today's dive," the Navy told the world, "is to demonstrate that the United States now possesses the capability for manned exploration of the sea down to the deepest part of its floor."

While it was true that the Navy, as claimed in the press release, had the capability for manned exploration down to the deepest of seafloors, it did not have the will.

Walsh and Piccard, along with their colleagues Rechnitzer and Shumaker, were summoned to Washington, D.C., in part so that President Dwight D. Eisenhower could hand out medals. *Trieste* was returned to her home port of San Diego. Soon after, someone in the chain of command—Don is not sure who—determined that *Trieste* should not dive below twenty thousand feet.

"I didn't agree with the change," Don tells me. "I knew the

engineering and design, and I've been in submarines and have a degree in engineering. But it wasn't that troubling because we weren't going to go back to 35,000 feet from San Diego. The likelihood of going back to the Challenger Deep was very, very remote. It would have required us to set up a base at Guam."

Don, recalling the order that restricted *Trieste*'s operating depth, pauses for a moment. "I have no hard feelings about it," he says. "It's just how it happened. The *Trieste* program was just fourteen people in the field. And after the dive, we thought we were going back, but it didn't work out that way. And by then the depth limitation was meaningless. *Trieste,* without a mother ship to support it, was not going back to the Challenger Deep."

On the *Trieste* dives, Don recognized something that others would see in the future with other submersibles, as small research submarines came to be known. He saw that the biggest expense was not in the submersible itself, but in the surface vessel needed for support. And that became one of his reasons for supporting AUVs. A futuristic AUV would not solve all undersea problems, but it might be capable of operating without tying up a surface vessel. It could work on its own, and it might even be possible to launch and recover it from shore. The support ship and all the people needed to run it might become, at some point, unnecessary.

Before a decade had passed, the Navy had more or less abandoned manned exploration of extreme depths, focusing instead on shallower parts of the deep sea, on regions of operational significance that could be cost-effectively militarized.

Fifty years later, in 2010, Don commented on the absence of a second dive to the bottom of the Challenger Deep. "It's amazing that no one has ever gone back," he said.

Two years after that, James Cameron, funded not by the Navy but by the National Geographic Society, Rolex, and others, dropped back down into the deepest abyss. Don, already elderly by normal standards, was on the support vessel. Just before the hatch closed, sealing Cameron into what could well be his tomb, Walsh offered advice. "I told him to have fun," Don says, "and to find that damn fish."

Cameron found no fish, but like Walsh and Piccard, he survived, and in all likelihood had what passes for fun in the minds of people happy to seal themselves into small metal canisters for seven-mile descents.

Be that as it may, no further missions were planned. And this, too, is part of the future story, of what is to come.

Don relates all of this to me toward the end of our four-hour conversation. We are wrapping up.

"How," I ask him, "would you define the deep sea?"

"There is no real definition," he says. "The words 'deep sea' convey an idea, a sense of other. It's deeper than your bathtub, but beyond that it's a fluid definition. It's more literary than actual. If you say you work in the hadal zone, people know that you mean depths below twenty thousand feet, but if you say you work in the deep sea, you are leaving the actual depth to interpretation."

The deep sea for *Trieste* was deeper than it is for a person submerging on a breath of air or in a helmet breathing unusual gases or aboard a military submarine or inside of any one of the world's many small submersibles, or even for an engineer controlling any existing AUV. But all of them operate beyond depths found in the bottom of a bathtub. All of them are part of the story of the deep sea.

And so my first question for Don Walsh remains relevant. What happened? What happened with exploration of the deep sea, however it might be defined, and what comes next?

Chapter 2

ON A BREATH

Free diving in Utila, Honduras, day ten—diving on a breath of air, without a pressure-proof sphere, without tanks or hoses to the surface, reliant only on my body and my wits. Free diving and still wondering how I am going to tell this story of people underwater, of humanity's ongoing relationship with the depths.

Wind ten to fifteen knots from the northeast, enough to stretch out a flag, enough to kick up a smart chop, enough to inconvenience *Trieste* if she were still in operation and not now a museum piece in the National Museum of the United States Navy.

Between dives, breathing, preparing for the next plunge, there comes an occasional splashing wave in the face, water interrupting the serious business of respiration. Then one last chestful of air followed by exhalation and then submergence.

This is not the Challenger Deep, but neither is it a bathtub. Is sixty-five feet deep enough? When diving with scuba gear or wearing a helmet, sixty-five feet seems shallow, but now, exhale diving, it is suddenly deep enough.

Exhale diving: I breathe systematically, filling my blood and tissues with oxygen, but rather than taking that last full breath along to the bottom—as any sensible person would do, as I myself would have done just a few weeks ago, before I undertook intensive training—I intentionally exhale. The lungs are usually an important reservoir for free divers, but exhale diving relies on oxygen stored in hemoglobin.

Deflated, I descend, something like *Trieste* after dumping gasoline. I sink not quite stonelike, but not like a diver clinging to two full lungs of air, forcibly submerging two balloons.

Exhale diving is a common training technique, and despite its undeniable misery, it is, oddly, fun. To sink through the sea, negatively buoyant, is pleasurable.

Some free divers claim that breath-hold diving and the oxygen deprivation that comes with it offer inspiration. Diving on a breath of air, they say, focuses the mind. I think that maybe it will help me see a way to tell this story of depth and submerged mobility and the capacity to accomplish meaningful tasks in an alien environment. Maybe a more focused mind will show me how to convert hundreds of pages of notes into an account that does not read like an extended and densely populated obituary appended to a series of accident reports.

And so I dive.

I descend through a sea that, but for the plankton, would be as blue as a glacial lake, an artist's blue, a perfect blue penetrated by sunshine. But the plankton smudges the underlying color, graying it, blotting streaks of light with tints of smog.

Descend headfirst, kicking but conscious of energy expended, aware of oxygen burned and carbon dioxide produced. Translucent plankton drifts past, jellyfish and ctenophores, but also beaded strings of colonial tunicates, the salps, and the so very numerous less identifiable creatures, some of the same living gelatinous stuff that Don Walsh and Jacques Piccard once saw through the viewports of *Trieste*.

Once or twice or three times each dive, something in the water, something not quite visible to aging eyes, stings exposed skin. Cheeks, lips, and wrists burn, a welcome distraction.

With exhale diving, even near the surface, even immediately after submerging, the subconscious brain calls out for air. That call is heard but purposely ignored.

Thirty feet down, the bottom rears up from shadows. Pay no heed to the urge to breathe, disregard that annoying suggestion begging for attention in the conscious brain, that braying donkey, the mind's irritating insistence that it might be best to turn around, to ascend, to surface. Focus instead on the smoggy blue. Pay attention instead to the plankton, to stings, to anything but this pesky yearning for a breath.

Continue downward.

At sixty-five feet, the remains of an eighteenth-century ship lie scattered on the seabed, mixed with coral and sponges and gorgonians, an octopus's garden. The wreckage was once the sailing vessel *Oliver,* carrying the remains of Central American trees, circa 1803, commanded by a man named Captain Hood.

Did sailors die here? Were men pulled down by the sinking ship, holding their breath, trapped belowdecks or tangled in lines? No one knows. All that remains are ballast stones, ribbing, and some sort of iron frame, rubbish but for what it once was, for its heritage, its reality. It is one of three million wrecks that litter the world's oceans.

Look around quickly. Take it in with a glance. Consider ending the dive, turning, heading upward. Pause to ogle the seascape once more. Calmly snub the voice inside requesting air. Block out the whines of senseless anxiety. Savor the appearance of Captain Hood's handiwork once more.

Now ascend. Push off the sandy bottom. Kick toward the surface.

No need to rush. Gaze down past my own long, graceful,

slow-moving fins while ballast stones fade from view. Focus again on the plankton. Listen to the noise of a passing boat, the *chug-chug-chug* of its diesel engine. Welcome all distractions, all things that keep the urge to breathe in perspective, in check, present but not overwhelming.

Sixty-two long seconds since submergence, arrive back home. Breathe wonderful air. Take one, two, three breaths, knowing that those first few inhalations on the surface truly matter, that they are what stands between consciousness and the threshold to something less than consciousness.

And then, those three breaths completed, signal the safety diver with the universal okay sign, thumb to forefinger. Signal that all is well, that there are no symptoms of an impending blackout. The brain, after this dive, will not succumb to low oxygen levels.

A shallow, short dive for an accomplished free diver, but exhaling before swimming to sixty-five feet presents a challenge for a post-middle-aged writer still learning about the sport, just now discovering what it is to control the body's need to breathe, just starting to appreciate in a very personal manner the meaning and reality of the mammalian dive reflex in human beings.

Another diver submerges, and I safety his ascent. And while a third trainee dives, I breathe up for another turn.

I submerge again, headed once more to the seabed.

Throughout the morning: breathe, exhale, submerge, touch bottom, return to the surface, inhale, recover, breathe some more, wait for my next turn. Repeat.

At the bottom, or sometimes on the way down, my diaphragm, in the parlance of free divers, contracts. From my instructor, an experienced free diver, a man who often swims deeper than two hundred feet on a breath of air: "Diaphragmatic contractions are to be expected."

Call them what you like, but they feel closer to convulsions than contractions. Gentle convulsions, but convulsions nevertheless.

Further words of wisdom based on his experience: "They are perfectly normal."

Normal, that is, if one chooses to hold one's breath long enough to raise carbon dioxide levels in the blood and for that carbon dioxide to lower the blood's pH, which in turn signals the brain that something is amiss. And the brain, for reasons known only to itself, signals the diaphragm, commanding it to contract, to put up a fuss, to convulse.

Let it. Carbon dioxide does not kill free divers. Low oxygen does. It is oxygen that keeps the machinery running. The brain senses carbon dioxide, which in the short term hardly matters, but the brain more or less ignores oxygen, which indisputably matters. A lot. Fair enough, high carbon dioxide usually means low oxygen. But the carbon dioxide response is alarmist. A few diaphragmatic convulsions — I mean, contractions — signify little of importance.

Free divers sometimes talk about oxygen saturation, which in this context is the level at which my blood and tissues are fully oxygenated. During a dive, as long as the oxygen level in my brain stays somewhere above seventy to seventy-five percent saturation, there is nothing much to worry about. Loss of consciousness is unlikely at levels above fifty-five percent saturation.

As the diving day continues, the thinking part of my brain turns meditative. Time wears on and the mind remains active but somehow changed as the body grows tired and the limbs heavy.

Robert Boyle occupies my thoughts. Boyle, the seventeenth-century alchemist, the experimenter, the scientist working fifteen decades before a largely forgotten man named William Whewell coined the word "scientist."

Boyle was not a diver himself, but his thoughts and experiments proved to be of great importance to diving.

In 1662, Boyle poured mercury into a U-shaped tube that was closed on one end. The weight of the mercury compressed air trapped at the closed end of the tube. Boyle used different amounts of mercury to create different air pressures at the end of his tube. He saw the air volume shrink and grow with the amount of mercury he poured in and drained out. Before long, he could measure the amount of mercury he used and calculate the volume of air that would be trapped, or he could measure the volume of air and calculate the amount of mercury.

Gas volume, Boyle realized, was inversely proportional to pressure. His realization became his law. Per Boyle's law, an increase in pressure reduces the volume of a gas. Decrease pressure, and the volume increases. It may be the first time in science that a physical law was expressed as an equation with two variables, one dependent on the other.

Boyle's law governs the lungs and the air they hold as a free diver submerges, or for that matter, the amount of air in a scuba tank, or the amount of air needed to purge the ballast tanks of a submarine at any particular depth.

On descent, I imagine my lungs shrinking, shriveling. On ascent, I envision them fattening, expanding, blossoming. Likewise, the air-filled passages leading to and from my lungs accommodate the rules of pressure and volume. I think of the same thing happening in diving whales, seals, otters, turtles, and birds.

Dive three times, four times, five times, and the mind moves toward a new place, slowing down, reining itself in. Day-to-day problems become less important. Thoughts, of course, drift in different ways for different divers.

While free diving, in keeping with the joking suggestion of my instructor, my coach, I sometimes compose and revise bad haiku. For example:

On descent lungs shrink
At depth wrung dry nothing left
Boyle stole my air.

Sometime around dive ten or twelve, my diaphragm comfortably convulsing on ascent, I realize that the story in possession of my attention is not about the achievements of one or two or three or even three hundred divers and submariners and engineers and physiologists. There have been government programs and corporate investments along the way, but there has been no NASA. Achievements below the waves cannot be attributed to a single program or even a handful of programs. There was no concerted push as there was for space. Cold War competition between nations played at most a minor role. There were few ticker-tape parades and no manufactured heroes.

While space exploration captured the public imagination and the public budget, undersea exploration struggled through the halls of academia, latched onto the needs of military and industrial operations, depended on the willingness of certain individuals to embrace great personal risk in complete solitude, and ultimately rallied around the billions of people who have now been underwater, down there poking around, whether on a breath or with a scuba tank or in a submarine or in front of a television screen. This is about achievements and experiences, large and small, from around the globe and across time. The story is not about one person or one group of people or one organization. It is the story of thousands of players often disconnected in time and geography and culture, men and women who not only did not know one another but who did not even know *of* one another.

I ascend from my last dive of the morning. I am breathless but unworried, my diaphragm jumping but my mind calm. I surface and inhale. I stare into the eyes of my safety diver and flash the okay sign, the silent "I'm just fine." There in the sunlit chop, after months of reading and interviewing important players and visiting submarines and manufacturers of what has come to be known as diving armor and inventors of underwater robots, the entire story comes together in my mind.

Free diving, breath-hold diving, is increasingly known as apnea diving. The word "apnea" is defined in the 2013 edition of *The American Heritage Dictionary* as the "temporary absence or voluntary cessation of breathing." It comes from the Greek *apnoia,* "without breathing."

Strangulation and choking cause apnea. Opiate overdoses result in apnea. Short bursts of apnea plague the sleep of millions of Americans. But apnea can also be voluntary. A highly trained apnea diver, or apneist, can cease breathing for some time and, on a single breath, reach great depths.

As of October 2016, the record voluntary breath-hold for an unassisted human, a person floating facedown in the water but retaining the ability to lift the head and breathe again without loss of consciousness, is an amazing 10 minutes and 49 seconds. Humans breathing pure oxygen immediately before voluntary apnea have lain facedown in water for longer than twenty minutes.

Not to disrespect these records, but the divers involved could accomplish very little aside from retaining consciousness while submerged. That is, on these record-setting dives, staying conscious and alive was enough. I mention that only as a reality check and to underscore a point that will become more important as the story of diving unfolds.

For other divers, including breath-hold divers working at

impressive depths, retention of consciousness is only part of the game.

~

Above a dock that pokes out into the clear water of Utila's only harbor, a sign proclaims SET YOURSELF FREE. Beneath those words ten free divers, ranging from rank beginners to Honduran and Austrian national champions, undertake two rounds of yogic breathing exercises. Most of the divers are in bathing suits. Some appear athletic, lean and muscled. Others less so, to a fault. Ages range from early twenties to seventy-two.

The first round of yogic breathing requires *uddiyana bandha,* or, in plain English, the upward abdominal lock. Exhale forcefully and then expand the rib cage as if inhaling, but without taking in any air. Done correctly, it leads to a freakishly deformed appearance that suggests the compression of the abdominal organs into the chest cavity. It trains the body for depths at which the lungs dramatically shrink. According to a yoga manual, the practice "stimulates the function of the pancreas and liver" and "removes lethargy," while also exciting something called "the digestive fire." For most of us, it is best described as decidedly uncomfortable. On the upside, it does not last long.

The second round requires *kapalbhati pranayama,* sometimes translated as "skull shining breath control." Exhale sharply and forcefully through the nose and then inhale passively but quickly. Repeat this rapid breathing cycle sixty times, take a couple of full breaths, and pause for sixty long seconds. Pull in a couple of recovery breaths. Do it all again.

The yoga manual says that *kapalbhati pranayama* "balances and strengthens the nervous system and tones the digestive organs." It also "removes sensory distractions" and "energizes the mind for mental work and removes sleepiness."

At first, the sixty-second pauses seem painfully long. With practice, they become contemplative and offer an opportunity to look around the dock, to watch the scuba divers next door loading tanks onto boats, to consider why a young woman sitting on the nearby seawall would cover her body with a collage of seemingly meaningless tattooed geometric shapes. The minute-long intervals also train the mind to accept the buildup of carbon dioxide. And they train the consciousness to be aware of the lungs.

The weight of the lungs varies from person to person, but in adult males they usually weigh just under three pounds, and in adult females somewhat less. The right lung is heavier than the left, which has to share space with the heart.

With inhalation, the diaphragm, a muscle, moves downward, giving the lungs room to expand. Air pushed by the pressure of the atmosphere passes through the nose and mouth into the trachea, which branches into bronchi and bronchioles in the lungs themselves, the plumbing dividing again and again before dead-ending into the microscopic sacs known as alveoli. Air rushes in, and the alveoli fill, expanding like the tiniest of balloons.

Inhaled air holds about twenty-one percent oxygen and less than one-tenth of a percent carbon dioxide. Oxygen moves across the expanding membranes of the alveoli into a mesh of tiny capillaries, then into the blood. At the same time, carbon dioxide moves in the other direction, out of the blood. In both cases, the gases are moving from places of higher concentration to places of lower concentration. Oxygen from fresh air moves from the lungs into the blood, while carbon dioxide generated by metabolism moves from the blood into the lungs.

During normal breathing, used air moving through the bronchioles and bronchi to the trachea and out through the nose and mouth is dramatically different from inhaled air. The twenty-one percent oxygen of inhaled air is replaced by the fifteen percent oxygen of exhaled air, now accompanied by six percent carbon dioxide.

Lungs should not be thought of as bags of air or as sponges. The divisions and the ever-smaller diameter of the trachea and bronchi and bronchioles allow the lungs to hold more than a thousand miles of airways. The smallest of these airways connect with something like 500 million alveoli. If the tiny air sacs were taken from the lungs of a single individual and stretched out flat, they could cover an entire tennis court. An appreciation of this tremendous surface area, this huge sheet of membranous tissue across which gases flow, is key to understanding breathing, and understanding breathing is key to understanding diving.

During a bout of *kapalbhati pranayama* breath-holding, during my contemplative sixty seconds, I think of those many miles of airways and how Boyle's law will shrink those 500 million tiny sacs during descent, and how that same law will see them grow during ascent. And I wonder what poor schmuck was talked into doing all the measuring and counting.

⌒

Another day of training, this time in deeper water.

One of the divers surfaces spitting blood. She had reached 150 feet, roughly the height of a fifteen-story building. At depth she had felt pain in her throat.

I swim over and ask to see the blood. She spits into her hand. The blood is in small but not tiny droplets suspended in her saliva. Probable diagnosis: throat squeeze. The plumbing above her lungs contracted to the point at which blood vessels burst.

A chest squeeze, which would occur deeper in the lungs and might rupture blood vessels in the alveoli themselves, would have given up finer droplets of blood, possibly brighter red in color, forming a pinkish foam in her spit.

She slams an open palm against the water surface, expressing frustration. The mind was strong and disciplined, pushing for depth, but in this case the flesh was weak.

In the world of free diving, a throat squeeze is no more than a minor issue, akin to, say, a twisted ankle in the world of soccer. She climbs into the boat and refrains from diving for two days.

Ears inconvenience divers. Anyone who has kicked down to the bottom of a swimming pool has felt the pain of water pressure on eardrums.

The eardrum—the tympanic membrane—separates the outside world and the outer ear from the middle ear. The eustachian tube, less than two inches long, connects the middle ear to the nasal cavities, which are in turn connected to the throat and the airways. If the eustachian tube were an open passageway, air would flow freely between the lungs and airways and the middle ear. But it is not an open passageway. Under normal conditions, it is more or less collapsed.

Swallowing can partly open the tube. Pinching off the nose, closing the mouth, and blowing gently outward can inflate it, forcing air through to the middle ear. Other tricks can be applied— the jaw can be wiggled, the head moved around on the neck, the seldom-thought-of muscles of the throat and mouth consciously tightened and loosened.

If a diver descends and air does not somehow pass through the eustachian tube to keep the pressure inside the middle ear equal to the pressure outside, Boyle prevails. The volume of the middle ear decreases. The eardrum flexes inward. Tissues around the middle ear flex inward or leak fluid. Pain results. The diver will either move air through the eustachian tube, return to the surface, or tear an opening through the eardrum.

Greek sponge divers of a bygone age often sacrificed their

eardrums, leaving their hearing impaired but ending any further need to equalize pressure through the eustachian tubes. Upon descent, their middle ears flooded. After surfacing, a sponge diver with perforated eardrums could, according to written descriptions, blow gently against a pinched-off nose to expel water from his ears, becoming for a moment a human fountain.

The sport of free diving is on the rise. Schools—usually an instructor or two and possibly a boat and a shoreside facility of some description—are popping up in the Caribbean, in Southeast Asia, in the Mediterranean. In the major urban centers of North America, free diving clubs seek time in swimming pools. In Russia, indoor free diving is well established. No one goes deep in pools, but they can go for distance in a discipline practitioners call "dynamic apnea." On a single breath, submerged just below the surface, people have swum the length of three football fields, nearly a fifth of a mile.

Regional and international contests in the open sea that focus exclusively on depth are beginning to attract attention. New records make the news. Deaths of champions attract headlines.

The emerging interest gives free diving the appearance of a new endeavor. It is not. Homer, Herodotus, Thucydides, and Plutarch wrote of free divers who could reach depths of two hundred feet. A guild of free divers known as the *corpus urinatorum* worked on bridges and piers in ancient Rome. Small populations—sponge divers from Symi in Greece, ama divers in Japan and haenyo divers in Korea, the Guna people of Panama's San Blas Islands—have been known as expert free divers for centuries.

A nineteenth-century Greek sponge diver working as his fore-fathers worked, without the benefit of a pressurized diving helmet and suit, lay on a plank balanced against the edge of a boat. In his hands, he grasped a large flat slab of marble that might weigh over a hundred pounds. A rope went from the stone to tenders on the boat. A second rope, lighter than the one attached to the stone but also managed by tenders, was tied to the diver's left wrist. The diver breathed. On a signal, his plank was tilted, sending the marble slab and the diver overboard and downward, trailing ropes behind. The diver wore neither fins nor mask. There were no depth gauges. Depths were estimated by the length of the line taken to the bottom. The marble slab provided both weight and a diving plane that could guide a rapid descent.

Unlike modern record seekers, the diver had work to do. On the bottom he harvested sponges. Soon he tugged the line tied to his wrist. Tenders on the surface pulled him, his marble slab, and his bounty upward and back to the world of breath.

The flat stone was called the *kampanelopetra*, the rope tied to his wrist was the *gassa*, the tenders were *kolaouzieris*, and the diver himself—the man at the end of the rope, the man holding the stone, the man working without air in the twilight depths—was the *voutto*.

In 1865, Royal Navy captain Thomas Spratt wrote about his observations of Greek sponge divers. He reported that some-thing like three thousand divers worked the sponge beds of the Mediterranean. He further reported that no more than five or six men died in diving accidents each year. "I have myself known many instances of divers going down to depths of from twenty to thirty fathoms," he wrote, "and I knew a family of three brothers, belonging to the island of Symi, who were called by their compatriots (and known to all the sponge-diving fraternity as) the Sarandaki, or the *Forties*, from their reputed capability of diving to that enormous depth. They were known to me more than twenty years ago, and for several years after; but only one of the three survives, and he is now employed in the Arsenal at Constantinople as the government diver. One of the other two lost his life whilst diving off the coast of Syria, either by a fit of apoplexy or by a fish, as the body was never recovered."

One fathom is six feet. The "Forties," if the stories are true, were reaching 240 feet.

And there is the documented 1913 dive of Stathis Hatzis, also known as George Kangis, George Chatzistathis, and Yorgos Haggi Statti. When an Italian battleship lost an anchor in deep water off Kárpathos, an island between Rhodes and Crete, its recovery became a matter of Italian military pride. Helmet divers, by then a common sight in Greece, could not reach the anchor. Stathis Hatzis, the locals told Italian naval commanders, could reach it on a breath of air.

Doctors examining Hatzis before the dive left behind notes. "Thorax circumference: 92 centimeters, 98 while deep breathing, and 80 while exhaling. Pulse: Between 80 and 90 per minute; from 20 to 22 respirations per minute. Weight: 60 kilograms. Height: 1.75 meters." In other words, Hatzis was a man of slight build with the pulmonary fitness of a smoker.

Examining his ears, the doctors discovered the ragged remains of a badly scarred and perforated eardrum on one side and no eardrum at all on the other. In addition, they noted, "he has pulmonary emphysema."

The doctors did not think that Hatzis should dive. Hatzis and those in charge ignored their opinion.

Recovery of the anchor called for more than a single dive. The anchor would first have to be located. Multiple dives, all deep and all long, would be required.

Accounts suggest that Hatzis attracted spectators. Sailors and local people would have watched him on the surface as he calmly prepared for his dive. Silently, he would have grasped his eighty-eight-pound marble slab, his *kampanelopetra*, and breathed for several minutes before each submergence.

Onlookers would have then seen Hatzis and his slab splash into the water, and some would have watched from directly above as he disappeared into the indigo depths. Even after his form dissolved in the blue, they would have observed tenders still slacking the ropes tied to Hatzis and his *kampanelopetra*. When he reached the seabed, they would have seen the ropes stop moving. And then they would have waited.

One minute would have passed. Then two.

For anyone waiting for a free diver, time passes slowly. Minutes are exceptionally long, the sort in which a wife sees herself as a widow. They leave spectators wondering if the man below should be referred to in the past tense. They allow the quick-witted to compose epitaphs.

And then the spectators would have seen the tug on the *gassa*, the rope joining Hatzis to his tenders. "Pull me up," those tugs would have demanded. "Bring me up now."

The dive that located the anchor lasted just under three minutes. The depth was recorded as 48 fathoms, or 288 feet.

Back on the surface, after recovering for a few minutes, Hatzis pinched his nose and blew water out through his ears.

In later decades, the great depths reached by Greek sponge divers were considered too deep to be credible. Stathis Hatzis was, temporarily, forgotten.

By the 1940s, free diving, which had for centuries been practiced as a profession in isolated pockets of the world, had become a form of recreation, a sport.

In 1949, just eleven years before *Trieste*'s dive to the bottom of the Challenger Deep, an Italian named Raimondo Bucher, an air force officer, bet fifty thousand lire that he could reach one hundred feet on a single breath. This would not be much of a feat for a Greek sponge diver or a modern-day trained free diver, but Bucher was neither of these things.

Bucher was warned that Boyle would not approve, that Boyle's law called for the brutally destructive crushing of his lungs. The dive, he was told, would kill him.

Bucher dived, surfaced with a smile, and collected his fifty thousand lire.

Over the next three decades, other free divers seeking to reach record depths emerged. Enzo Maiorca and Jacques Mayol famously competed, for the most part against each other, diving beyond three hundred feet. By the 1960s, women were diving for records. In the 1980s, Enzo Maiorca's daughters, Rossana and Patrizia, reached depths well below two hundred feet, and Angela Bandini passed three hundred feet. Tanya Streeter, riding a weighted sled down and a balloon back up, passed 525 feet to claim a record previously held by a man. Russian Natalia Molchanova consistently broke records well into middle age. Despite many hours in the water laboriously perfecting her own form and training others, she often dived for the sheer joy of it.

In 2015, off the east coast of Spain, engaged in what amounted to playful training, Molchanova disappeared, never to be seen again, presumed dead at age fifty-three. Free diving, although a pastime, was and is a serious business.

~

Mayol, the Maiorcas, Bandini, Streeter, Molchanova, and Hatzis, it turns out, were unlikely to perish at the deepest depths of their dives. Few free diving fatalities occur on the bottom. Those who push themselves too far lose consciousness on the way up or on the surface itself and die as they slowly sink back downward.

At depth, the lungs are compressed. Oxygen is concentrated in the blood and tissues. The pressure of the oxygen can be measured in millibars, or thousandths of an atmosphere. Consciousness requires about twenty-seven millibars of oxygen pressure in the brain, which requires about forty millibars of oxygen in the lungs. Neither the numbers nor the units are especially important. What is important is a visualization of the lungs and blood and tissue during ascent.

Boyle showed that decreasing pressure increases volume and increasing volume decreases pressure. During ascent, the lungs expand. As the pressure in the lungs drops, oxygen still held in the tissues and blood—already at lower levels than normal this far into the dive—moves into the lungs. It is as if the lungs, on ascent, suddenly suck up the body's remaining oxygen. The brain, abruptly oxygen starved, falters. The diver may or may not notice the fading of consciousness, and even for those who are aware of their impending doom, there is little to be done. It is not as though the diver has the option of stopping, of ceasing the ascent to avoid unconsciousness. The surface beckons.

Among survivors, the transition to unconsciousness is seldom remembered. Their rescuers have to explain to them that they dropped out, faded away, fell victim to a shallow-water blackout, to what a medical doctor might call cerebral hypoxia.

And the reality is that most victims of a shallow-water blackout are survivors, as long as a rescuer is at hand.

That rescuer is, typically, the safety diver. Free divers should and almost always do work in pairs or in teams, one diver down and one or more divers up. Those on the surface wait and watch, ready to act. Usually, the safety diver submerges to accompany the ascending diver along the last thirty feet or more of a deep dive, swimming next to the ascending diver through what is sometimes called the danger zone or the death zone, ready to seize his or her ward if and when the ascending diver checks out, ready to carry the semiconscious or unconscious diver to the surface, prepared to remove the mask and noseclip, ready to tap the victim's cheeks or to blow into the eyes or to puff a rescue breath directly into the mouth, ready to calmly but firmly spit out the command "Breathe!" In most cases, consciousness returns within seconds.

Or a diver might be conscious all the way to the surface, where he or she exhales, decreasing an already low oxygen pressure in the lungs while at the same time, by virtue of the effort involved in exhaling, increasing oxygen demand in the body. An immediate inhalation delivers oxygen to the blood, but that blood will take some time to reach the brain. The conscious brain, losing patience, shuts down.

Loss of consciousness may not be complete. A diver close to the edge might experience nothing more than what has been called loss of motor control, often shortened to LMC, in which the head might jerk backward or the arms or legs might shudder or lurch. The safety diver, again, pulls away the mask, taps the cheeks, blows into the eyes. The victim in this case is conscious but groggy, with mental integrity compromised, the brain drifting in that netherworld familiar to a patient coming off anesthesia or a child waking from a deep sleep.

In the lexicon of free diving, a loss of motor control event is a samba. The characteristic backward jerking of the head reminded someone, somewhere, of the Brazilian dance. The term stuck.

To competitive free divers, the samba is not unusual. Just as serious weight lifters and runners tolerate muscle aches, competitive free divers accept sambas and even shallow-water blackouts. As long as a rescuer is on hand and attentive, small things like loss of motor control or temporary loss of consciousness will not keep them in the shallows.

My free diving fins, long and narrow, at one time said COMPETITIVE FREE DIVER in white letters against a black background.

But understanding sambas and shallow-water blackouts, I had no intention of becoming such a diver. With a scrap of sandpaper, I scrubbed away the "com" and "itive," reducing myself to the status of "pet free diver." Routine acceptance of the loss of motor control and consciousness is not in my game plan.

From a song recorded by jazz legend Ella Fitzgerald, in reference to the dance but with relevance to free divers:

I only dance samba
Go, go, go, go, go

From the earliest times, free divers trained other free divers. In Greece fathers coached sons, and in Japan mothers educated daughters. The International Association for the Development of Apnea was organized in 1992. Performance Freediving International and Apnea Total came later, as did branches of organizations originally established for scuba divers. The world's largest scuba organization, the Professional Association of Diving Instructors, known at dive shops around the world as PADI, recently entered the free diving arena. PADI's entrance brings the sport into the mainstream diving world. With its seal of approval, free divers will no longer be seen as possibly suicidal renegades in a world dominated by scuba.

It is hard to imagine PADI, an organization very focused on safety, accepting the routine spitting up of blood and the chasing of records that involve dives so deep and so long as to ensure occasional loss of consciousness and, now and then, death. Nevertheless, PADI is now part of the sport. If it stays in the game, it may become a restraining influence, just as it was for deep scuba diving.

Whether or not fair-minded restraint is welcome in the free diving world remains to be seen.

~~

Back offshore, day eighteen of free diving, with slow-moving three-foot swells from the east, rising and falling, rising and falling. We are tied to a mooring close to shore. We hear tumbling waves crashing against a low limestone bluff that was once a coral reef. There is, too, the noise of our own boat slapping against the water.

The depth drops quickly from the shoreline. Above shallow sand the water is turquoise blue, above seagrass a darker blue, above corals and sponges a mixed ever-changing blue, beneath us the light-swallowing blue of substantial depth.

I slip over the side of the boat, pull my "pet free diver" fins onto my feet, and swim the short distance to our dive buoy. Facedown, breathing through a snorkel, grasping the buoy, which bounces up and down with the swells, I feel for the pulse in my neck.

My resting heart rate, ashore, hovers in the mid-fifties. Here, just after reaching the buoy, not rested, I count seventy beats in a minute.

To prepare, I remove my mask, forcefully exhale, and then pull myself down along a weighted line. Removing my mask, allowing seawater to contact my face and eyes, accelerates my warm-up and the triggering of my mammalian dive reflex. I return to the surface, breathe for a minute or two, empty my lungs, and pull myself back down. I repeat the forced exhales five times, never dropping below thirty feet. Then I exhale passively, without vigor, and visit sixty feet.

Now ready, I pull my mask over my face. I relax on the

surface, breathing through my snorkel, which allows me to lie facedown, lazily conforming to the water's movement. The motion disturbs but does not destroy the peace.

After five minutes, I check my pulse. Sixty-four beats per minute. I fill my lungs, not exhaling at all, and pull myself down the weighted line. I move willfully and painfully slowly. There is no need to hurry here. At thirty feet I stop, holding the line with my left hand and reaching for my neck with my right hand. In one minute, I feel sixty beats.

I surface and repeat the exercise. I feel fifty-four beats.

On the fourth try, I go a bit deeper, to thirty-six feet. In one minute, I count forty-eight beats.

The fifth dive, to thirty-seven feet, also yields a pulse of forty-eight.

The sixth dive, now at forty feet, gives me a mere forty-two beats.

I am far from the first to recognize that something happens to a diver's body in the water. Between 1890 and 1894, French physiologist Charles Robert Richet, who would eventually win the Nobel Prize for unrelated research, strangled ducks by tying a cord around their necks. He strangled some on the surface, and he strangled some underwater. The average duck out of the water died after seven minutes, but the average submerged duck lasted twenty-three minutes. In submerged ducks, heart rates dropped. Oxygen was conserved. Something about the water seemed to trigger a physiological response.

Later, Swedish-born Per Scholander, trained as a medical doctor but interested in botany, traveled to the Alaskan Arctic. When not looking at plants, he watched seals diving. Ringed seals are common in the Alaskan Arctic. Adults typically weigh a little more than a hundred pounds. They dive during the warm season, when the sea is not covered by ice, but also during the cold season, under the ice, periodically popping up through holes, seemingly unconcerned about the possibility of drowning, of being trapped under the frozen surface. Dive times can exceed twelve minutes.

Scholander noticed that seals exhaled before they dived. Taking measurements on captive seals, he found that their heart rates dropped when they dropped underwater. He found, too, that the blood flow to skeletal muscles was, more or less, cut off. During a dive, large muscles functioned without oxygen, through anaerobic respiration, producing lactic acid that created what came to be known as an oxygen debt, a debt that had to be repaid after surfacing.

During a dive, the blood that normally flowed to the muscles was diverted to the nervous system. When it came to oxygen, the nervous system exercised the privileges of rank.

Scholander moved from seals to porpoises, ducks, penguins, and a beaver. He and others followed with otters and muskrats, with mammals of all sorts, with various other birds. The patterns were the same.

And there was more. There was a detectable contraction of the spleen, and with it a sudden release of red blood cells that increased the body's capacity to carry oxygen. And there was a measurable slowing of brain activity, a drop in the rate at which synapses between neurons fired, a change believed by some free divers to explain the odd meditative state that accompanies their dives.

Scholander and other researchers switched to humans. Volunteers, holding their breath in pools and pressurized chambers and at times in seawater, were instrumented, wired up like patients in intensive care. They were sent to the bottom to exercise vigorously, but even then their heart rates slowed. In humans as in other animals, what came to be known as the mammalian dive reflex, or simply the dive reflex, or what Scholander himself once called the Master Switch of Life, accompanied submergence.

More than a century after Richet's strangled ducks and seventy years after Bucher's dive on a fifty-thousand-lira bet, in a world where organized schools teach ordinary people to reach what were once considered not only astounding but fatal depths, sentences like this one, from the *Journal of Applied Physiology,* are common: "The diving response, exhibited by all air-breathing vertebrates, is elicited by apnea and consists of peripheral vasoconstriction due to sympathetic activity, connected with initial hypertension, and a vagally induced bradycardia with reduction of the cardiac output."

My pulse, in dive seven, comes in at forty-five beats per minute. Other divers, fitter than me, more dive adapted, have reported pulse rates below fourteen beats per minute, levels that should not, according to physiologists, support consciousness in humans. One possible explanation: their heart rates were higher than fourteen beats per minute, but the blood vessels feeding extremities were so constricted as to render the pulse at the wrist and fingertips difficult or impossible to detect. A second possible explanation: the physiologists are wrong.

In seals, heart rates might drop from more than one hundred beats per minute on the surface to as low as ten beats per minute at depth. Ten beats per minute—barely alive but in fact doing just fine, swimming long distances, jetting ahead and then suddenly pausing, hunting, bellowing out strange whistles and calls, staring into cold blue water with impossibly engaging brown saucer eyes, to all appearances frolicking beneath the waves.

⌐

Breathe normally. Each inhale and each exhale carry about a quart of air, the tidal volume of lungs at rest. Now pull in a slow, deep, deliberate breath, expanding your abdomen and then your rib cage and then your chest, doing your level best to truly fill your lungs. If you are a typical man, you now hold about six quarts of air, and if you are a woman, a little more than four quarts.

Now exhale forcefully, expelling everything to the best of your ability. Your lungs still hold about a quart of air, your residual volume.

Take another slow, deep, deliberate, chest-filling breath, hold it, and submerge. Those six or four quarts of air are buoyant. They provide around twelve pounds of flotation for a man and eight for a woman. But never mind. Submerge anyway.

Thanks to Boyle, at thirty-three feet, lung volume is down by fifty percent. Keep going.

Somewhere around one hundred feet, your lungs are no longer anything like full. In fact, they are almost as empty as you can make them by forcefully exhaling on the surface. They have compressed to something approaching residual volume.

Keep going. You may begin to feel uncomfortable, as if your chest is being squeezed. And, of course, it is. The weight of seawater is pushing in from all sides. But continue downward anyway, knowing that you may damage your lungs or your airways. That damage may be nothing more than mild pain and coughed-up blood, but it may also be far worse. Never mind. Chase those few extra feet, that personal-best depth, or, if you are good and from a relatively small nation, push for that national record. Then experience a full lung squeeze, a lung squeeze bad enough to fill your alveoli, at least some of them, with fluid, with blood.

Back on the surface, when you breathe, you cannot catch your breath. Your lips turn slightly blue. Your blood-filled, fluid-flooded alveoli are of little use. They give you an idea of what it is like to suffer from pneumonia or asthma. Someone may hand you an oxygen mask, which might keep you alive long enough to recover.

An alternative approach to extreme depths involves more patience. Build up to that record descent, diving daily or almost daily, picking up a few extra feet each week. If it is difficult, stop, turn around, go back to the surface. Give your body time to adapt. It seems that, over time, the diaphragm becomes more flexible. Vascularization—the degree to which blood vessels fill the lungs—may increase. The lungs can shrink well below their size following a forced exhale without pushing blood into the alveoli. At least, that is what seems to be happening. That is why the world champion free divers, the men and women capable of depths beyond three hundred feet, work for years not only on their technique but on their bodies, on their lungs, on their chests. Some of them may not look like athletes, but they are. To compete at these levels, to reach these depths, they were probably born with certain physiological advantages, but training has definitely extended their natural abilities.

For many years, the competitive free diving community operated as if the sport's only real risk was loss of motor control or loss of consciousness, usually near the surface. Well-trained safety divers controlled that risk.

Enter actor and competitive free diver Nick Mevoli, the first American to swim from the surface to 330 feet down and back on a breath of air. Unlike that of many top competitors, his free diving career had a steep trajectory. He entered the sport somewhat late in life. Natural talent, self-confidence, and drive compensated for missing years.

In competitions, divers announce their discipline, their anticipated time, and their planned depth. A diver might, for example, choose to compete in "free immersion," meaning that he or she will pull downward and back upward on a vertical line without actually swimming. Or a diver might choose to compete in "constant weight no fins," meaning that he or she will swim without fins and without the aid of a line. The announced time is a matter of convenience, something watched by safety divers awaiting the hopeful competitor's return from the dark abyss. A tag will be set at the announced depth.

The competitive diver has to retrieve the tag, show it to the judges on the surface, and follow a strict recovery protocol. A diver returning to the surface tagless will be awarded no credit for depth. Likewise, a diver losing muscle control or consciousness near or on the surface, with or without a tag, will receive no credit.

On November 17, 2013, in a competition in the Bahamas, Nick Mevoli announced his intent to swim to 236 feet without fins. He had been deeper, but always with fins. Swimming downward and back upward with bare feet requires far more oxygen. The discipline known as "constant weight no fins" is considered more challenging than the disciplines that allow the use of fins or lines or weights that can be dropped at the bottom.

Nick took his breaths. He submerged. Sonar tracked his descent. An announcer called out his depth to spectators on a nearby beach and in boats and even in the water, looking downward.

At about 223 feet, the sonar detected a pause in Nick's descent. Competitive divers, realizing that things are not going as planned, often turn around early. The announcer told the crowd that Nick might be ending the dive early. But this was not correct. Nick had simply paused to change position, from headfirst to feetfirst, an old trick that helps clear pressure in the ears. He continued his descent.

At 236 feet, his target depth, he lost more time struggling to find the tag. He began his ascent. The announcer, apparently concerned, told the safety divers to be ready.

Three minutes and thirty-eight seconds passed before Nick surfaced, almost a minute longer than planned. He flashed the okay sign, but then struggled. The judges waited for the three words required by the protocol: "I am okay." Nick mumbled incoherently. He failed the surface protocol. He clung to a safety line, fighting to breathe. Almost a minute passed before he fell backward, unconscious.

The safety divers and other medical personnel tried but failed to coax him back to awareness. They lifted his still unconscious body onto a floating platform. One of the safety divers saw blood dripping from his mouth, trickling from his crushed lungs up and out and into the ocean.

Adrenaline was administered by injection, but still Nick remained unconscious.

Twenty minutes passed before he was carried ashore and rushed to a nearby health center. His vital signs were gone. He was declared dead, the first fatality ever to occur during an organized free diving competition.

Later, someone asked one of the doctors familiar with Nick's death if it would be fair to say that he had drowned in his own blood. The doctor preferred medical terminology, but she did not disagree. Nick's alveoli, squeezed beyond endurance, had flooded. Surrounded by air and by medical responders, Nick had drowned in his own blood less than two years after his entry into the world of competitive free diving.

Ben Franklin, in a time before swimming became popular, was a swimmer. He was also the inventor of the precursor of swim fins: he wore wooden paddles strapped to his hands. And he was the inventor of bifocals. My dive mask is outfitted with what are, in effect, bifocals. Without them, I could not read the diving computer on my wrist that tells me how deep I am and how long I have been underwater.

Today is Ben Franklin's birthday, and I am in the water celebrating his many accomplishments. For three minutes, I breathe systematically on the surface. Four deep, slow breaths, each taking ten or twelve seconds, first filling the lower abdomen, then expanding the ribs, then inflating the chest, and then relaxing my shoulders to feed air into the uppermost part of my lungs. Then exhaling slowly, through pursed lips. Then three more relaxed breaths, quicker. Then four more deep and slow, followed by three more relaxed, repeating the cycle until I am ready, or until I am simply bored by the process, or until one of my dive partners grows impatient. At that point, I follow the three breaths with one more slow, deep inhale, one more exhale, and a final inhale. I tuck and dive, folding at the waist, head down, stroking with my arms, then sending my feet upward above the water so that their weight in air boosts my downward trajectory.

In my own admittedly backward salute to Ben, I wear no fins. I am diving as Nick Mevoli was diving on the day of his death, in the discipline of "constant weight no fins." Unlike Nick, I cannot take my free diving seriously. I call my dive a Tarzan dive because the stroke is similar to that used by the jungle man when swimming in crocodile-infested rivers. It resembles a frog stroke, arms reaching out beyond the head, sweeping back for thrust, legs sprawled outward and then jerked together, neck straight with the top of the head facing downward for streamlining. This is not about looking around. This is about depth.

A single stroke near the surface, where my lungs remain full and buoyant, is good for maybe eight feet. In front of me, through blue water, I focus on a vertical white line, a line tied to weights that dangle 101 feet below, deeper than I intend to go. I am too old for foolishness.

At thirty feet, I swim through a reverse thermocline. The deeper water is suddenly warmer than the surface water, the effect of some sort of upwelling or of rain cooling the shallows.

My lungs, compressed by the weight of water, have shrunk. A single stroke propels me fifteen feet downward.

Below the thermocline, the water is clearer, with less plankton. The artist's brush has applied a darker hue. I can hear, with increasing clarity, the sound of a distant ship.

At sixty feet, my lungs are smaller still. I am sinking now, no longer buoyant. A stroke speeds my descent but is not entirely necessary.

I have never before Tarzan dived below sixty feet. But I feel good today. I continue downward.

To conserve oxygen, I pause between strokes. I pause and glide and sink and think. I watch the white line against the darkening background.

And then, to my surprise, right there in front of my nose, I see the weights at the end of the line. I touch them. I have reached 101 feet. With one hand, I grip the line, arresting my fall and turning my body around, following the protocol of the discipline.

My strokes begin again, but without the energy I felt at the surface. On upward stroke number one, my limbs feel full of syrup, of lactic acid molasses.

My diaphragm contracts. The feeling is less than comfortable but not alarming. It is familiar. Carbon dioxide makes me want to breathe but falls far short of making me need to breathe.

Upward stroke number two, and I have already returned to eighty-five feet, but I still fight my own weight. I pay now for the free ride to depth, for my negative buoyancy.

Another contraction. Unnervingly strong.

Two more strokes to reach sixty feet. The water's blue is noticeably lighter. My body, too, is noticeably lighter, my lungs expanding, regaining buoyancy. My urge to breathe grows. The syrup in my limbs remains sluggish, but my brain retains control. It demands measured, deliberate strokes. There is no need to rush.

I pass thirty feet. I am above the reverse thermocline, back in cooler water. I am fully buoyant. I can once again pause between strokes. I pause and glide and float upward. I relax and try to enjoy the diaphragmatic contractions.

And then the surface. The air. I exhale hard and then inhale deeply.

But wait. I have pushed myself too hard. Very suddenly, I feel myself fading.

A moment later, I am back in the world.

A samba was never in my game plan, but nevertheless I just had my first. The conscious part of my brain, alarmed by the lack of oxygen, took a little rest. I lost a moment during which my safety diver gripped my neck to keep my head above water, removed my mask, tapped my face, and ordered me to breathe.

I am told later that I uttered the word "samba" just as I dropped out. Most people do not say "samba" as awareness falters. I have no idea what, if anything, this suggests about the inner workings of my mind. I myself have no memory whatsoever of the event.

Happy birthday, Ben Franklin.

With fins, after four weeks of training, I reach a depth of 132 feet. I surface without drama, without a samba. Not so deep at all for a competitive free diver, nor for someone breathing a pressurized mixture of helium and oxygen, nor for a submarine, and especially not for *Trieste,* but I surface thinking about Don Walsh's words defining the deep sea not as a place but as a state of mind. On a breath, I am in the deep sea at 132 feet. And back on the surface, I realize that I have lost sight of my free diving goal. I will never set depth records. I will never swim past 400 feet, as Alexey Molchanov did in 2018. I will never Tarzan dive past 330 feet, as William Trubridge did in 2016. My goal is to be comfortable at around 60 feet, at depths that competitive free divers would consider suitable only for warming up.

The free diving community has been criticized for its focus on depth. Free divers, some say, just go down a line, grab a tag at a targeted depth, and return to the surface. They see very little along the way. Serious competitors do not even wear masks, but instead dive with noseclips and bare eyes or flooded goggles. The point is not to see but to be, not to look at the reef or the sand or the kelp or the fish, but merely to be at depth on a single breath, apneic, chasing a record. There is, with a few exceptions, no real question of accomplishing meaningful work in today's free diving world. Even in the distant past, with a few exceptions—notably the Greek sponge divers, but others as well—something more than free diving was needed if meaningful work was to be accomplished.

Enter the humble bucket. Not the ordinary humble bucket, but a large, heavily weighted bucket. Turn it upside down. Sink it, with air trapped inside. Boyle mercilessly compresses that air, shrinks its volume. A diver standing in the bucket would feel water rising over the feet, up to the waist, even wetting the chest. The diver might bend over inside the bucket, grasping objects on the seabed. And from there it would be a small step for the diver to duck out of the bucket, working on the bottom for a minute or so before popping back into the bucket, into his or her own personal bubble, for a breath. The diver could pop in and out, again and again, until the air became stale, full of carbon dioxide and depleted of oxygen.

This is not at all a hypothetical bucket. Aristotle wrote of just such a bucket in the fourth century BCE, and Alexander the Great supposedly descended in what amounted to a barrel made from glass during the 332 BCE Siege of Tyre.

The bucket became known as a "diving bell." And an actual bell—a church bell—with its weight, size, and shape, might in fact make an excellent rudimentary diving bell.

After Aristotle, the diving bell appeared sporadically through history, described here and there in various accounts. In some harbors, diving bells were a common sight.

Edmond Halley, already known for mathematically predicting the return of the comet that would henceforth bear his name, already known for his treatise on trade winds, wrote of diving bells in his wonderfully titled "The Art of Living under Water: Or, a Discourse concerning the Means of furnishing Air at the Bottom of the Sea, in any ordinary Depths," published in the *Philosophical Transactions,* dated 1714–1716. About the bells that were in his time seen in harbors and bays throughout the world, he wrote, "The diver is safely conveyed into any reasonable depth, and may stay more or less time under water, according as the bell is of greater or lesser capacity" and "at thirty three foot deep or thereabouts, the bell will be half full of water."

The problem with bells of the day, Halley recognized, was twofold. First, all that water entering the bell as depth increased left the diver wet, and the diver, as Halley put it, "will not be long able to endure the cold." Second, the bell had to be repeatedly raised to renew the air, to replace the bell's trapped air bubble as respiration removed oxygen and added carbon dioxide.

Halley built his own bell from wood. It measured three feet across at the top and five feet across at the bottom, with a height of around five feet. "This," he wrote, "I coated with lead." A glass window at the top of the bell lit his little world.

So far his bell was more or less like others in use. But then he added barrels, "about 36 gallons each," also covered with lead and, like the bell, open at the bottom. A hose—a leather "trunk" coated with oil and beeswax to render it waterproof—was attached to each barrel. "Thus I effected," he wrote, "by a contrivance so easy, that it may be wondered it should not have been thought of sooner, and capable of furnishing air at the bottom of the sea in any quantity desired."

Imagine his bell being lowered from the side of a ship. Imagine, too, his barrels going over the side. The hoses from the barrels led into the bell. When the barrels were lowered to a depth slightly beyond that of the bell, air would run through the hoses into the bell, renewing the diver's atmosphere. By means of a rope run from the surface to Halley's hand, he could signal to have barrels sent down or retrieved, to have the bell lowered or raised. And if line signals proved inadequate, he wrote notes on sheets of lead with the aid of an iron stylus.

Halley himself dove, he wrote, with as many as five others, in water as deep as sixty feet. He stayed, he claimed, for longer than an hour and a half at depth, sending barrels back and forth to the surface. By lowering the bell slowly and adding air from barrels as he descended, he could keep pace with the changing pressure, allowing him to stay entirely dry. But divers, he noted, could exit and reenter the bell. He pointed toward potential applications: pearl diving, sponging, bridge work, harbor work.

Free diving, Halley and others before him saw, had limitations. To accomplish complex tasks, divers needed more time at depth. They needed to be able to work. Working divers, for many tasks, have to breathe under water. Despite their many faults, compressed gases, whether sent down in barrels or by other means, are part of the story.

Chapter 3

UNDER PRESSURE

Between 1913, when Stathis Hatzis held his breath to recover an Italian anchor, and the 1970s, the world of diving changed. To leap ahead without, for the moment, thinking about the six decades of intervening innovations, consider diving in 1971. That year, just over a decade after *Trieste*'s dive to the bottom of the Challenger Deep, when the official free diving record was a mere 250 feet, the American submarine USS *Halibut*—the second Navy submarine of that name—carried a team of highly trained divers far into the Soviet Union's territorial seas. The submarine was, of course, nuclear-powered.

U.S. Navy submarines often penetrated into Soviet waters. And some, like the *Halibut,* carried a strange protuberance, an odd hump standing out from the deck. In other American submarines, the tacked-on bulge was a deep submergence rescue vehicle, or DSRV, designed to rescue crews trapped on the seabed. The thing on the *Halibut*'s deck looked like a DSRV, but it was, in fact, a saturation diving system. The technology was a critical component in the top secret and perhaps oddly named Operation Ivy Bells.

Four hundred feet beneath the surface of the Soviet Union's Sea of Okhotsk, working from the swollen bulgy thing that looked like a DSRV but was not, divers onboard the *Halibut* opened an air-lock door and exited into frigid northern waters. These were not free divers or ordinary scuba divers. They were saturation divers, using recently proven techniques that allowed them to stay at depths of hundreds of feet for days or weeks on end, eating and sleeping in a pressurized chamber when they were not out in the ocean working.

When in the water, they were attached to the saturation system and the submarine by umbilicals. On their backs, they probably wore rebreathers, backpacks designed to recycle the exotic mix of helium and oxygen that they breathed, made in such a way that helium was conserved and that no telltale trail of bubbles was left behind. The word "probably" is appropriate because many of the operation's details remain shrouded in secrecy. But other details are public knowledge. For example, it is known that they could talk to colleagues in the submarine, their words carried through wires in the umbilicals. And hot-water suits—wet suits lined with perforated tubes that bathed the divers in heated water pumped through the umbilicals—kept them warm in seawater that hovered around forty degrees Fahrenheit.

Their task in the Sea of Okhotsk, their job in their futuristic undersea space suits there in the shadow of the hulking *Halibut,* was to locate and tap the telephone cable that linked Vladivostok's fleet headquarters with a submarine base in Petropavlovsk-Kamchatsky, home for Soviet Delta and Yankee class ballistic missile submarines.

Operation Ivy Bells was so secret that most of the crew aboard the *Halibut* did not even know that the divers were tapping a telephone cable that buzzed with information about Soviet submarine activity. The crew, for the most part, thought the divers were recovering pieces of supersonic anti-ship missiles called SS-N-12 Sandboxes that the Soviets had fired into their Sea of Okhotsk test range. The missile recovery work itself was, of course, top secret, as the Soviets would not have approved of their Cold War nemesis recovering missiles that belonged to them and that were resting on the seabed within their own territory. So, from the almost-too-strange-to-be-fiction department, the ultrasecret Operation Ivy Bells had a cover story that was itself top secret.

The clandestine tapping of a subsea telephone cable is not a five-minute job. The divers did not just go in and out. The tap necessitated hours of work, and it required maintenance. In the early phase of the operation, recordings had to be retrieved each month. Technology improved, and the tap had to be replaced. Navy divers, living under pressure, went back again and again. And when they were not tied up with the telephone tapping operation, they were busy shoring up their top secret cover story, recovering bits and pieces of missiles.

Were it not for a disgruntled and financially strapped U.S. government employee named Ronald Pelton who worked with the recordings, U.S. Navy divers might still be visiting the tap today. But Pelton sold information to the Soviets, including a description of Ivy Bells. Eventually, Pelton went to prison, and the tap, retrieved by the Soviets, went to a museum in Moscow. It was twenty feet long and weighed three tons. Scattered publicly available reports claim that it was powered by plutonium. For the Soviets, during a time of détente, of Strategic Arms Limitations Talks, of a general easing of Cold War tensions, here was clear evidence of American misbehavior.

Putting cloaks and daggers aside, ignoring international intrigue and national security and naval heroics and ultimate betrayal, one might view Ivy Bells as something of a diving miracle. Saturation diving, a technique that had been entirely experimental ten years earlier, had become, for the USS *Halibut* divers and for commercial divers working in the world's growing offshore oil fields, routine.

To understand how this minor miracle came to be, how a form of diving that was not only unavailable but almost unheard-of and undreamt-of less than half a century earlier, requires a little background. Context is likewise required to understand its limitations.

~~~

In 1913, when Stathis Hatzis recovered the Italian anchor by repeatedly free diving to 288 feet, helmet divers were a common sight within the Greek sponge diving community. The Greek divers had known of diving helmets since at least 1863, when a man named Fotis Mastoridis provided a demonstration in a Greek harbor.

The free divers dived on a breath of air, but the helmet divers breathed underwater, inside their helmets. They breathed compressed air pumped down from above.

Helmet diving met some resistance in Greece, but by 1867 at least twenty helmets were in use on the sponge beds. The Greek sponge diving community became divided between those who used helmets and those who held their breath. Helmet users tended to become rich faster, but they also died with far greater regularity. An 1867 report tracked twenty-four helmet divers, claiming an astounding ten fatalities by the end of the year.

The death rate might suggest that the helmet was a new invention, a buggy device full of dangerous uncertainties. In fact, early versions of the traditional deep sea diver helmet and suit—the copper helmet and breastplate bolted to the rubberized canvas suit with attached lead boots—had been in use since the 1830s. John and Charles Deane of Kent, England, are sometimes recognized as the inventors. Originally, the contraption was used as a smoke hood, a means for ducking in and out of smoke-filled ship holds at a time when shipboard fires were all too common. But the brothers quickly realized that their smoke hood, weighted to overcome its buoyancy, could work underwater.

75

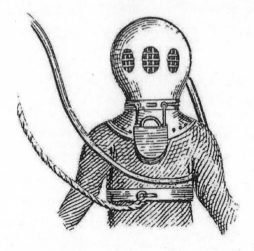

One version of the diving helmet invented by John and Charles Deane. The brothers realized that their helmet, originally intended for use in smoke-filled rooms, could also be used underwater. What looks like a padlock around the diver's neck is a weight, needed to counteract the buoyancy of the helmet. The air supply hose runs under the diver's arm, preventing tenders on the surface from putting stress directly on the helmet's hose fitting.

Envision, more or less, a bucket with windows and an air hose. Air pumped into the helmet gave the diver his own little private atmosphere. As new air was pumped in, bubbles flowed out from the bottom of the helmet. If the diver stood upright, his entire head and neck stayed dry. Bending over released air and let water into the helmet. The Deanes soon added a "skirt" or "jacket" to the bottom of their helmet, allowing divers to lean forward or sideways or backward without flooding. The helmet and jacket would be recognizable to working divers today, and many of them would unhesitatingly agree to perform light work in the gear, in part because of nostalgia but also because the design was then and remains now entirely functional.

By the 1830s, Augustus Siebe, working with the Deane brothers and others, was marketing a diving helmet with a complete suit attached. In other words, Siebe extended the skirt used by the Deanes into a set of waterproof coveralls with built-in boots and rubber seals around the wrists. Because air could no longer flow out through the bottom of the helmet, exhaust valves were added. Air flowed in through the hose from the surface and out through a valve controlled by the diver. If the diver closed the valve, the suit ballooned, carrying the diver upward. If the diver opened the exhaust valve, air flowed out and the suit compressed around his body.

British engineer Augustus Siebe modified and improved on the earlier helmet designed by the Deane brothers. Helmets similar to those made by Siebe, seen in this illustration (titled Divers Preparing for Work, *which appeared on the front cover of the February 6, 1873, issue of the* Illustrated London News*), continue to be used in some parts of the world.*

The suit that Siebe made—which was, in fact, based on the gear made by the Deanes, along with improvements crafted by Siebe himself as well as others—would be modified again and again and again, but in its many forms it would continue to be used by working divers into the 1970s and beyond. It was known generically as "standard dress" and "heavy gear," but specific brands and models carried names and nicknames such as the Mark V, the Morse, the Chinese 3 bolt, the Desco 3 light, the Yokohama, and the Toa. Regardless of the name, Siebe would recognize all of them.

So why were all those helmeted sponge divers dying? And why, in 1913, when diving helmets and suits that would be used almost unchanged into the 1970s were readily available, did the Italian navy turn to a free diver, a breath-hold diver, to recover its anchor from a depth of 288 feet?

Picture yourself as a helmet diver in 1913. At that time in Greece, it probably was not possible to reach 288 feet in a helmet because available air compressors simply could not provide air with enough pressure to reach that depth. But for the moment suspend disbelief. Imagine that you are a helmet diver going to 288 feet in 1913, sent there to find an anchor. You are unusually knowledgeable, somehow at least familiar not only with your equipment and the sea around you but with some of the information coming from researchers, from scientists.

You slip into your rubberized canvas suit, putting your feet through the wide collar opening and then pulling it on like loose-fitting but very stiff coveralls. Your feet end up nestled inside the suit's attached socks, and your hands poke through wrist openings with rubber seals.

You sit on a stool while two tenders attach boots over your suit's socks, boots that go on something like a pair of sandals but that each weigh ten pounds. Next, the tenders lower a copper breast-plate over your head and onto your shoulders. The breastplate holds an additional thirty pounds of lead weights. The tenders work the wide collar of your rubberized canvas suit around the outside edge of the breastplate, slipping the breastplate's twelve threaded studs through matching holes around the edge of the suit's collar. Then they slide four elongated plates over the studs and tighten nuts to hold everything in place. With that completed, your breastplate is sealed to your suit. Your head sticks up through the middle of the breastplate, making you look, for the moment, something like a turtle.

The tenders wrap a weighted belt around your stomach. It holds forty pounds of lead. Suspenders cross from the front of your belt over your shoulders to the back of your belt, supporting the weight.

But for your helmet, you are dressed. Your job so far has been to sit tight, appearing stoic and maybe contemplative, refusing to acknowledge the discomfort of the breastplate against your collarbones.

If you have an urge to urinate, it will have to wait.

If you smoke—and if you are a helmet diver in 1913, chances are good that you do—it is at this point that one of the tenders inserts a lit cigarette between your lips. While you smoke, a lifeline is tied to your breastplate. The lifeline is nothing more than a piece of rope, probably hemp or manila, that will join you to the surface. In a few minutes, you will rely on that rope to communicate with your tenders by using sharp tugs.

You exhale the last drag off that cigarette, and your tenders lower a copper helmet over your head. They spin it one-quarter turn to engage it with the neck of the breastplate, and you are sealed in. You can hear air hissing into the helmet, entering from the hose attached to its lower left edge, far enough down and back that it is not spitting into your ears or across your eyes.

Including your lead, you are now wearing more than a hundred pounds of gear. You are ready to dive.

Nearby, two men turn the handles that power an air compressor. There is no diesel engine attached to the compressor. There is no electric motor. By hand, the tenders spin the crankshaft that translates their effort to the pistons that push air into the hose attached to your helmet. You feel the air against the back of your neck.

The helmet you are using has not one but two valves, the first for controlling air coming into the suit and the second for controlling the exhausted air. You can adjust the incoming flow with the valve on the side of the helmet. You close the exhaust valve, allowing the incoming air to partially inflate the suit, and then you open the exhaust valve just a turn or two, balancing the incoming and exhausting air. The helmet restricts your vision, and you cannot see either of the valves, but you know where to find them. You know their location as well as you know the location of your nose and ears.

With the help of two tenders, each lifting under one arm while also steadying you, you stand. Assisted, you take two or three steps to reach the edge of the boat. You could walk on your own, barely, but on the moving deck of a boat, dressed in a manner that would make any fall awkward at best, you accept the help.

Then you take another guided step from the edge of the boat onto a platform hanging over the side.

The tenders lower the platform with you on it. Your lifeline and your hose, managed by the tenders on deck, hang in the water beside you. As the platform meets the water, you feel the pressure against the outside of your suit. If your tenders were quick, if all actions were performed efficiently, perhaps fifteen minutes have passed since you slipped into the rubberized canvas suit up there on deck, back there in the world. And now the dive begins. You watch the sea move across your helmet's faceplate. And then you are completely submerged. You are underwater. The terrible weight of the lead and the suit disappears, countered by the ocean's buoyancy.

Twenty feet down, below the worst of the sloshing caused by the waves, you step off the edge of the platform, holding your lifeline with one hand. You adjust your valves carefully. If you open the exhaust valve too far, you will sink quickly, and as you sink, the pressure will increase, further reducing the volume of your suit and accelerating your descent. If you descend too fast, the suit will push against you with the weight of the water above your head. You have to control your descent, allowing the compressed air to keep your suit inflated against the pressure of the sea. You may not be formally trained in Boyle's law, but you can feel it at work. You do not want to sink like a stone. You want to sink slowly, in a controlled manner.

Your tenders slowly play out the lifeline and hose. As you descend, your ears feel the pressure. You yawn or wiggle your jaw, opening your eustachian tubes, equalizing the pressure inside and outside.

You pass sixty feet, seventy feet, ninety feet.

If your hose breaks somewhere above you and there is no one-way valve between your helmet and the hose to prevent air from escaping upward, you will become suddenly intimate with Boyle's law. Your suit will collapse around you with remarkable force. Wrinkles and creases of rubberized canvas will pinch and bruise your flesh. If you are deep enough, you will suffer far more than pinches and bruises.

A typical male's body has about 3,000 square inches of surface area. At 99 feet, there is about 45 pounds per square inch of water pressure on your body. Do the math. That's something like 135,000 pounds of pressure—the weight of, say, a small ship.

Put another way, if your hose breaks, reducing the pressure inside your helmet to that of the surface, the sea will attempt to stuff your body up inside your helmet. Even with a small change in pressure, the kind that might occur if you suddenly fall thirty feet, the sea will squeeze blood and fluids from your limbs and torso into your head, the only part of your body protected from the outside by a rigid surface, by your helmet, leaving you with bloodshot eyes and bruises on your legs, arms, and torso.

You have heard stories of divers being crushed into their helmets, bodies turned to jam by the pressure of the water, and while you know these stories to be exaggerations, you are very much aware of the stark reality of water pressure.

But all of this is manageable. Your tenders know how to manage your lifeline and hose. You know how to manage your valves. You will descend slowly and your compressor will keep up. Boyle will not kill you today.

Somewhere around 150 feet, you begin to relax. The world, it seems to you, is a funny place. You also notice that your valve adjustments are becoming increasingly clumsy. Here, at 150 feet, you feel the effects of nitrogen under pressure, of narcosis, sometimes called rapture of the deep. It has been compared to the effects of rapidly consumed martinis and to the effects of laughing gas, but neither comparison is quite right. Nitrogen narcosis—first described by scientists in 1834, not long after the invention of your helmet and suit—has its own unique impact on the workings of the brain.

The early researchers attributed symptoms to changes in blood flow under pressure, but by 1899 researchers realized that the symptoms were somehow linked to nitrogen's solubility in fat. Cell membranes, including those of brain cells, are fatty. Dissolved nitrogen molecules bind to receptors on brain cells.

Even today, the exact mechanism behind nitrogen narcosis remains cloudy, but the effects are well known. Euphoria is accompanied by irrelevant fixations, a weakened ability to multitask, a decreased ability to solve mathematical problems and to pass memory tests, and, sometimes, laughter. Also, the water does not seem quite as cold as it should. Nitrogen narcosis warms you, or, more accurately, tricks you with a false but welcome feeling of warmth.

You fool with your valves but continue downward. This is not your first deepwater rodeo. You know enough to concentrate but not fixate, to ignore but enjoy the euphoria, to remain diligently mindful.

At 240 feet, you are entering the zone in which euphoria, in some divers, turns to terror. Around this depth or a little deeper, deteriorating cognitive ability can give way, in some divers, to hallucinations. Some have reported hearing and even seeing trains. Continue downward and stupefaction becomes inevitable.

But you are not going much deeper, and, again, you have been here before. You know that every dive is different, that the effects of narcosis can vary from dive to dive almost as much as from diver to diver. You know that your past history of diving this deep does not protect you on every dive, that this one may be different. But still, you are confident. You are contained. You continue downward.

Now, at 288 feet, your lead boots land on a mercifully firm seabed. You are stoned, completely bombed, but you hold on to the presence of mind needed to balance the flow of air coming into and going out of your suit, maintaining slightly negative buoyancy. You moonwalk slowly on the bottom of the sea, stirring up clouds of silt. From within your private narcotic haze, you recall that you are here to find a lost anchor.

But you have another problem. The nitrogen has made you stupid, but not so stupid as to forget the possibility that the oxygen in your air, now entering your tissues at a heightened pressure, can cause convulsions and loss of consciousness without warning.

This particular type of oxygen toxicity—there is another that is less relevant to this kind of diving, to short-duration dives—has been known since 1878, when Paul Bert recognized that it could affect insects, spiders, birds, and humans. It is sometimes called the Paul Bert effect, as if Bert invented it, when in fact you cannot blame him for oxygen toxicity any more than you can blame Boyle for the nefarious problems of pressure and volume.

Some divers who have succumbed to oxygen toxicity and survived report tunnel vision or tingling lips or twitching or ringing in the ears before convulsions and loss of consciousness, but for the majority of divers the worst symptoms are entirely insidious. Muscle contractions followed by spasms strike without warning, bewitching the body and sending it into the throes of what medical professionals recognize as a tonic-clonic seizure. Loss of consciousness follows quickly on the heels of seizures.

The seabed is not the place to experience such symptoms. In most cases, oxygen seizures underwater are not accompanied by survival. But at this depth, at 288 feet, you should be reasonably safe from oxygen toxicity. In all probability, you are not quite deep enough to succumb to a tonic-clonic seizure.

You have another problem. This deep, your tenders cannot deliver all the air that you would like to have. As you descend, more force is needed to squeeze the compressor's pistons down. With every foot of descent, the tenders work a little harder. To avoid exhaustion, they take turns. At times, as many as four men struggle at the compressor, two on each side, working short shifts, frequently replaced by new tenders, by fresh muscles.

But still, the flow through your helmet slows. The air in your little world gets a bit stale, a little stuffy. Carbon dioxide levels build, carrying additional risks, among other things increasing the symptoms of narcosis and the chances of an oxygen seizure. You are careful to avoid exertion. This is not the time for huffing and puffing.

Maybe you find the anchor, or maybe you find nothing but sand and fish and sponges. Either way, you finish your business and leave the bottom. Your tenders bring you up slowly. Almost immediately, the risk of oxygen toxicity goes from low to nonexistent. As you ascend, you may not remember every aspect of your dive, but the nitrogen narcosis fades and your mind clears. So far, neither Boyle nor Bert has killed you. You have once more bluffed your way through the narcotic effects of nitrogen at depth. But you cannot go directly to the surface. In 1913 in Greece, you do not know the details, but you have heard that you have to ascend slowly or stop along the way. The nitrogen that dissolves in fat cells also dissolves in the blood that circulates through your veins and arteries and that permeates your tissues. Think of it like the bubbles in a bottle of champagne. Unopened, the cork holds the pressure inside the bottle, and gas dissolved in the champagne stays in solution, invisible and harmless. But unleash that cork, remove the pressure that holds the gas in solution, and the liquid turns to a bubbling froth, pouring out of the top of the bottle, spilling all over the floor. Unless you are suicidal, you do not want your blood and tissues to bubble like a suddenly opened bottle of champagne.

Even a few bubbles are entirely unwelcome. Even if you ascend slowly enough to avoid turning your blood into a bubbling froth, the formation of one or two or a few small bubbles in the wrong place, in your arteries and tissues, will be a problem. Surface too quickly and suffer from decompression sickness, or bends, with symptoms coming on as you pass through shallow water or even hours after your dive, when you are relaxing over a cup of coffee, perhaps telling sea stories to your tenders. You have seen it in others, in divers who developed strange itchy rashes, who bent over in pain so severe that it brought on tears, who could not talk, who staggered across the deck, who complained of nauseating dizziness, who could not walk at all, who could not see or urinate. And you have witnessed it in divers who simply died, sometimes before their helmets were removed from their heads.

You yourself, and every helmet diver you have known, have felt joint pain—agony in the knees and elbows and hips, pain that sometimes resolved itself after a few days or a few weeks or a few months, but also pain that persisted indefinitely.

And so, no matter how cold it is, no matter how eager you are to get to the surface, to have that helmet removed from your head, to have that breastplate unbolted, to shed those ten-pound boots, to peel off that rubberized canvas suit, the need to proceed slowly to the surface is paramount. That much you know.

Understanding why it is paramount and knowing how slow the ascent should be took years of work by a number of scientists. In that quest, certain names stand out. Those names include Dr. Andrew Smith, Paul Bert, and, most of all, John Scott Haldane.

Few divers understood decompression sickness in 1913, but that is not to say that decompression sickness was not understood, at least in part.

Robert Boyle himself, two and a half centuries earlier, put animals under pressure and studied the results. He noticed bubbles in the eye of a snake—a "viper," as he called it—that he had, out of curiosity, pressurized and depressurized. "The little bubbles," he wrote, "by choking up some passages, vitiating the figure of others, disturbe or hinder the due circulation of blood."

In 1840, Charles Pasley, involved with the salvage of the HMS *Royal George,* which went a long way toward promoting the use of Siebe's deep sea diving helmet and suit, noted that none of his divers "escaped the repeated attacks of rheumatism and cold."

It was not the diving itself that brought on Pasley's "rheumatism." Men often worked in environments belowground where pressurized air could prevent flooding by groundwater. These men were called caisson workers—"caisson," from the French word meaning "box," in this context referring to a pressurized container. They suffered in much the same way as Pasley's divers.

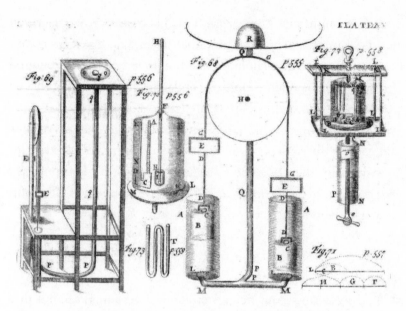

Robert Boyle's physical experiments with gas pressures and gas volumes led to what is known today as Boyle's law. Boyle generated both low pressures and high pressures using various devices, including this one, which he referred to as "our engine." (The Philosophical Works of the Honourable Robert Boyle Esq.: Abridged, Methodized, and Disposed under the General Heads of Physics, Statics, Pneumatics, Natural History, Chymistry, and Medicine)

In 1870, 352 caisson workers were digging the holes for the foundations—what engineers know as the footings—that would support the Eads Bridge connecting St. Louis, Missouri, and East St. Louis, Illinois. Each shift, the workers would climb down a spiral staircase into the ever-deepening shafts they were digging. They would pass through an air lock, a sealed steel cylinder six feet in diameter and six feet tall, with pressure-proof doors on both sides. Going down, they would enter the air lock through the north door, close the door behind them, and open a valve, allowing pressurized air to fill the lock. Once the lock reached the same pressure as the caisson below, they would open the second door, the east door, and proceed into the caisson. Inside, at the work site, they would dig. They might work knee-deep in groundwater, but the pressurized air held the worst of the flooding at bay.

At the end of their shift, they would return to the air lock through the east door, close it, and open a valve to bleed air from the lock, lowering its pressure back to one atmosphere. And then they would exit through the north door and climb the spiral staircase to the outside, to sunlight, to fresh air.

Eventually, the diggers would reach bedrock, where the bridge footings would find firm ground. But over time, as they dug deeper into the earth, the pressure had to be increased to keep the caissons dry. One caisson was pressurized to fifty pounds per square inch, the equivalent of the pressure found on a dive to about one hundred feet.

As the caissons grew deeper, workers complained of joint pain and temporary paralysis. Men reported stomachaches, difficulty urinating, pus in their urine, loss of bowel control, severe pain, and other complaints. On March 19, worker James Riley came out of the air lock, reported that he was okay, and then fell over. He was dead fifteen minutes later. Another worker died the next day. Three days passed, and three more men were dead.

Many workers experienced various symptoms, most of which simply went away. Others were never affected. But before the bridge was completed, twelve men died and thirty were seriously injured. No one knew what was happening, but some started calling it "bridge disease."

While work progressed, physician Alphonse Jaminet investigated. Inside the caisson, he heated drinking water and river water and alcohol and ethanol, noting the boiling points. He took pulses, including his own. He noted that voices changed in the caisson.

Dr. Jaminet, like the workers, experienced bridge disease first-hand. In a written report, he described the symptoms, referring to himself as "the author," in the third person. "Ten minutes after returning to the normal atmosphere," he wrote, "the author commenced to feel a severe epigastric pain, which was relieved by taking a table-spoonful of a cordial; on going home the pain left him completely, but there was a general feeling of great fatigue, which lasted about three hours."

Writing of a subsequent caisson visit and another all-too-personal encounter with bridge disease, he switched to the first person. "Both legs and my left arm were paralyzed, still I was suffering in both with excruciating pains, which I can only compare to pains felt after a fracture of the left leg, which I experienced some years ago." He treated himself with rest and regular spoonfuls of "Jamaican rum." His knowledge of what had happened to some of the workers, along with the tone of his account, leaves little doubt that the possibility of his own death crossed his mind, but he recovered, more or less, within a day.

In his report, he noted that none of the men were affected under pressure, "but always after returning into the air-lock, or going out of the air-lock and returning to the normal atmosphere." He noted, too, that the amount of pressure made a difference. At low pressures, no one suffered at all. Nevertheless, he had no idea what caused the symptoms. "No precise or even proximate cause of the pathological phenomena," he wrote, "based on facts, could be adduced, nor was any prophylactic means advised in order to avoid the recurrence of such cases." He set up a hospital on-site to treat symptoms, but the treatments involved little more than pain relievers and beef broth.

At about the same time, hundreds of miles away, construction of the Brooklyn Bridge had six hundred caisson workers digging out the footings on either side of the East River. Just as was true at the Eads Bridge, every downward foot required additional air pressure to hold back the groundwater. On the Brooklyn side, the excavation ended at forty-five feet with a pressure of seventeen pounds per square inch, and injuries were rare. On the

Manhattan side, the diggers reached forty-five feet in a month. At any one time, there could be as many as 125 men under pressure, working eight hours each day. As they dug deeper, shifts were shortened. Below forty-five feet they worked seven and one-half hours, and below fifty feet they worked seven hours. By then, most of the workers were complaining about symptoms of one kind or another. Unusual and even debilitating fatigue was among them. An elevator was installed so that the workers would not have to climb steps at the end of a long shift.

At fifty-one feet, Dr. Andrew Smith joined the project. He had been an army doctor and was currently a surgeon and a throat specialist. Now he was surgeon to the New York Bridge Company, tasked with figuring out what was wrong with these caisson workers. Like Dr. Jaminet in St. Louis, he did not understand what was happening, but he issued commonsense rules of the apple-a-day variety. The men should not go under pressure with an empty stomach, they should pursue a meat diet, they should drink lots of coffee but little or no alcohol, and they should limit exercise immediately after leaving the caisson.

Smith examined employees and job applicants. Those with obvious signs of alcoholism, heart disease, or lung disease were dismissed. Those accepted for work in the Brooklyn Bridge caissons ranged in age from eighteen to fifty. Many lived in tenements with perhaps twenty or thirty roommates.

The men lacked faith in Smith's recommendations. They believed in particular that he was mistaken about alcohol. Smith himself administered alcohol to those in pain, and what worked as a treatment, some of them reasoned, must also work as a preventative.

"The habits of many of the men were doubtless not favorable to health," wrote Smith, "but everything which admonition could do, was done to restrain them from excesses."

Smith, among others, including Jaminet, noticed that the men could not whistle under pressure. "The utmost efforts of the expiratory muscles is not sufficient to increase materially the density of the air in the cavity of the mouth, hence on its escape there is not sufficient expansion to produce a musical note," he wrote. He noticed the same change in voices that Jaminet had reported, and it seemed to him that the workers breathed both more rapidly and more deeply than they would outside, back in the world. He found body temperatures were one or two degrees above normal. He recorded pulse rates that quickened immediately after pressurization but that slowly dropped through the day until they became "small, hard, and wiry."

While the men worked, Smith experimented with dogs and pigeons, learning nothing.

The men, although poor, were New Yorkers. They knew of the strange posture then fashionable among wealthy women, the bent-over walk called "the Grecian Bend." It reminded them of themselves, their posture altered by pain and paralysis. So rather than picking up Jaminet's name for their symptoms, bridge disease, they referred to their affliction as "bends." Smith preferred the more formal and descriptive "caisson disease."

A man named Frank Harris later wrote of his experience in the caissons of the Brooklyn Bridge. He claimed to earn more in one day than he was paid on the surface in two weeks. He described pressurization and ear pain and burst eardrums. Once inside the caisson, he wrote, work began. "The sweat was pouring from us, and all the while we were standing in icy water that was only kept from rising by the terrific pressure." He wrote of blinding headaches. And this: "In the bare shed where we got ready, the men told me no one could do the work for long without getting the bends." He claimed to see a colleague who had recently returned from the caisson collapse to the ground, wriggling in a struggle with pain. "The bends," he wrote, "were a sort of convulsive fit that twisted one's body like a knot and often made you an invalid for life." Men who appeared perfectly healthy might suddenly stumble and fall, sometimes overcome by agony and other times simply numbed and crippled.

Smith recorded 110 cases of caisson disease that required treatment. From his case notes: "Case 12—Joseph Brown, foreman, American, aged about 28. Taken on the 28th of February, about an hour after coming up from a three hours' watch. Excessive pain in left shoulder and arm, coming on suddenly, 'like the thrust of a knife.' Pain continued until he went down again for the afternoon watch." Another man suffered from "dimness of sight and partial

The caisson workers who built bridge footings in the nineteenth century were often bent over in pain from decompression sickness. They sometimes called this sickness "bends," after the popular stooped posture known as "the Grecian Bend," considered fashionable in some urban centers at the time. The term "bends" is still used by divers throughout the English-speaking world. This lithograph by Thomas Worth (published by Currier & Ives in 1868) caricatures the Grecian Bend.

unconsciousness." Another "was taken with numbness and loss of power in the right side, also dizziness and vomiting."

Treatment involved doses of ergot, ginger, or whiskey, as well as hot baths, poultices, ice packs, injections of atropine, and sometimes morphine. There were many cases that did not require treatment and even more that went unreported by men who were eager to hold a highly paid job, who equated masculinity with indestructibility, and who knew that they could possibly recover, or not, with or without treatment.

Just as in St. Louis, some men died. The first was a German, about forty years old, who collapsed well after his shift ended when he was climbing the steps of his boardinghouse. Eight days later, an Irishman, about fifty years old, collapsed in the air lock, succumbing to the strange disease before he even reached the surface. He regained consciousness. He begged for water. He suffered from convulsions. And soon after, he was dead. Another man returned to the surface vomiting, then collapsed, and then suffered through the night before death put an end to his misery.

Among the victims of caisson disease was the man in charge of the bridge, Washington A. Roebling. He was a hands-on engineer, a man no more hesitant to work at great heights than to descend into the caissons. Often, he would be back and forth to the caissons several times in a day.

After ascending from what was no more than a routine visit to the job site, no different from dozens of others, he collapsed in pain. His family and close associates did what they could to downplay the importance of the incident, fearing that it might add to the growing sense of dread in the workforce, scare investors, and undercut the success of the Brooklyn Bridge itself.

Roebling's own account describes a treatment that involved "stupefaction by morphine, taken for twenty-four hours internally until the pains abated." His wife privately remembered that she did not expect him to live through the night. As was the case for many who suffered from caisson disease, Roebling's symptoms subsided over time. But he was among the less fortunate survivors, those who never fully recovered. Pain and numbness came and went, uninvited and unexpectedly. Dizziness and vomiting stalked him through the years. Caisson disease—bends, decompression sickness—had ruined another man's life.

Roebling wrote, "The labor below is always attended with a certain amount of risk to life and health, and those who face it daily are therefore deserving of more than ordinary credit." He was referring to the workers in the caissons, but also to himself.

Like Alphonse Jaminet, Smith recognized what by then must have been obvious to everyone. Caisson disease did not occur under pressure, but rather after the pressure was released. Pasley's "rheumatism" was not caused by diving itself, and it was not caused by pressure itself, but rather by the reduction of pressure. The men suffered from the effects of returning to the surface. It was not the depths but the surface that was hurting and crippling and killing.

Smith and the workers also recognized another strange facet of caisson disease. A victim, if not too badly affected, might go back under pressure to find relief. "It frequently happened under my observation," Smith wrote, "that pains not sufficiently severe to deter men from returning to work were promptly dissipated on entering the caisson, to return again on coming into the open air."

Putting the pieces together, Smith believed that the root cause of caisson disease was the distribution and redistribution of blood in the body. Under pressure, blood was squeezed into the body's core, "from the surface to the center," as he put it. Upon ascent, the blood moved out of the core, but the blood vessels were slow to respond, resulting in "disturbance of function," in caisson disease, in bends.

This belief was utter nonsense.

But in one very important regard, Smith was entirely correct: "It is altogether probable that if sufficient time were allowed for passing through the lock the disease would never occur." That is, caisson disease could be avoided by slow depressurization of the air lock that joined the pressurized work space to the outside world. Although he knew that the duration of a shift was an important factor, he focused on the magnitude of the pressure in the caisson, what to a diver would be depth. To further complicate matters, he believed—correctly, to some degree—that each person had different susceptibilities, that one man's adequately slow ascent would leave another man doubled over in pain. But he had no way to predict how much time was needed, how slow was slow enough for the average man, and he was aware of the costs of "sufficient time," the time during which men would have to be paid even though they were not actually working.

Smith suggested five minutes ascending in the air lock for every thirty-three feet of caisson depth, but the workers, eager to be on their way after their long shifts belowground, tended to ignore the suggestion, just as many of them ignored his suggestions about drinking.

Although Smith never prescribed recompression as a means of treatment, he had seen men relieved by it. He had watched workers go back into the caissons in pain, only to find relief with depth. With that observation, he suggested construction of what would become known as a recompression chamber, a cylinder in which victims of caisson disease could be repressurized. "Let there be constructed of iron of sufficient thickness," he wrote, "a tube 9 feet long and 3½ in diameter, having one end permanently closed, and the other provided with a door opening inward, and closing airtight." The tube would sit horizontally, and patients on a bed or stretcher would be slid into the chamber. The door would be closed and the pressure slowly increased. Once symptoms disappeared, the victim would be slowly brought to the surface.

"By occupying several hours, if necessary, in the reduction of pressure," Smith suggested, "it is probable that a return of the pain could be avoided."

Smith did not understand the cause of caisson disease, but he had a good notion of the cure. More than a century later, recompression in a chamber remains the only effective treatment available for serious symptoms of decompression sickness.

⌒

Paul Bert, the Frenchman who had recognized nitrogen narcosis, also worked on caisson disease. Among other things, Bert put twenty-four dogs in a pressure chamber and took them to a depth of 290 feet, then brought them back to the surface in less than four minutes. Twenty-one of the dogs died. Only one showed no symptoms at all. He exposed more-fortunate dogs to the same pressure but brought them up gradually, over an hour or longer. They, of course, survived.

Bert realized that gas bubbles—which formed when compressed gas came out of solution during what amounted to an ascent from depth—were killing the dogs. He also discovered that the bubbles were full of nitrogen. While the dogs were breathing pressurized air, nitrogen dissolved in the blood and tissues. Bert knew of the narcotic effect of nitrogen under higher pressures, but under relatively low pressures—pressures too low to trigger dangerous symptoms of narcosis—nitrogen in itself was harmless. During a gradual ascent, the nitrogen could slowly leave the tissues and blood to be exhaled through the lungs. Bubbles in meaningful quantities and of a size sufficient to cause symptoms would not form. But during a rapid ascent, the dissolved nitrogen turned to very troublesome bubbles. It did not matter whether those breathing the pressurized air during the ascent were caisson workers or divers, dogs or humans.

A bubble blocking blood flow to a joint might cause pain. A bubble in the brain or a bubble blocking blood flow to the brain might cause paralysis or death. One did not require the intellect of Paul Bert to understand that bubbles in the blood and tissues, especially those large enough to block capillaries and arteries and veins or to press against nerves, were not conducive to good health and comfort.

Bert, unlike Smith, understood the cause of caisson disease, but his solution was the same as that suggested by Smith. Caisson workers and divers must decompress slowly, "they must not only allow time for the nitrogen of the blood to escape but also to allow the nitrogen of the tissues time to pass into the blood." Like Smith, he saw the value of recompressing victims. And he saw the value of breathing pure oxygen as a means of treatment. Nitrogen bubbles, his thinking went, would disappear more quickly in victims breathing oxygen. He was right on all counts.

By 1890, during the construction of the Hudson River Tunnel, air locks were used to recompress injured caisson workers. By 1900, Augustus Siebe's company, the diving suit manufacturer, was funding a British physiologist named Leonard Hill, who experimented with frogs to gain insights into how quickly people could safely return from a particular depth to the surface. And, above all others, there was the Scotsman John Scott Haldane, a luxuriantly mustachioed scientist known for his work on the physiology of breathing and his work in mines.

Haldane had a reputation for experimenting on himself to determine the effects of various gases and their absence. For example, in one set of experiments, he locked himself inside a sealed box six feet long and four feet wide and high until outside observers, watching him through a window, saw him vomiting and opened the box. Some of these experiments ran longer than seven hours.

The decompression chamber used by John Scott Haldane and his colleagues in the development of the world's first decompression tables. The chamber, made from riveted plates, leaked constantly. This and many other interesting photographs are included in the landmark paper "The Prevention of Compressed-Air Illness," published in 1908 by Haldane and his colleagues in the Journal of Hygiene. (*The Prevention of Compressed-Air Illness,* A. E. Boycott, G. C. C. Damant, J. S. Haldane J Hyg [Lond] 1908 Jun; 8[3]: Plate V, PMCID: PMC2167126)

In 1905, the Royal Navy came to Haldane with questions about decompression. While Haldane experimented with goats, he might have preferred to use baboons or chimpanzees, which had more in common with humans but were expensive and more difficult to coax into a pressure chamber. Goats were inexpensive, cooperative, and at least somewhat similar to humans in their size and physiology. "Goats," Haldane wrote, "while they are not perhaps such delicate indicators as monkeys or dogs, and though they are somewhat stupid and definitely insensitive to pain, are capable of entering into emotional relationships with their surroundings, animate and inanimate, of a kind sufficiently nice to enable those who are familiar with them to detect slight abnormalities with a fair degree of certainty." In other words, he would be able to observe signs of decompression sickness in goats, especially if they were not distracted by food or goats of the opposite sex. Even trivial signs of discomfort could be observed if the goats were allowed, he wrote, "to fall into a state of meditative boredom."

Tables and text from his reports describe the various symptoms he observed in goats. Often, it might be nothing more than a goat mouthing an injured limb. A black-and-white photograph published in one of his papers shows a goat holding a foreleg off the ground, as if lame. Some goats were crippled, both temporarily and permanently. "Goat 9," he wrote, "had a curious short seizure and rolled over on the ground 10 minutes after decompression." Many of the goats did not survive his experiments.

Haldane also experimented on himself, his son, and other human volunteers. His son, who grew up to become a scientist, would later write of the perforation of eardrums that sometimes occurred during their work in pressurized chambers: "The drum generally heals up; and if a hole remains in it, although one is somewhat deaf, one can blow tobacco smoke out of the ear in question, which is a social accomplishment."

Haldane's work showed that both goats and caisson workers could be held at a depth of 33 feet indefinitely without apparent problems. Thirty-three feet has the pressure of two atmospheres—the first from the weight of the atmosphere itself and the second from the weight of 33 feet of seawater. Haldane reasoned that if diving goats and people could go from two atmospheres to one atmosphere without trouble, they might be able to go from six atmospheres, or 165 feet, to three atmospheres, or 66 feet, without trouble. And they could, usually.

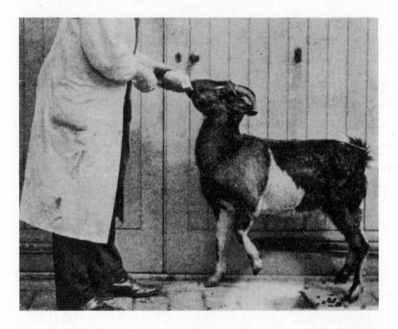

One of Haldane's experimental goats showing signs of decompression sickness, according to Haldane, in its right foreleg. This photo is from "The Prevention of Compressed-Air Illness." (The Prevention of Compressed-Air Illness, A. E. Boycott, G. C. C. Damant, J. S. Haldane J Hyg [Lond] 1908 Jun; 8[3]: Plate V, PMCID: PMC2167126)

Led by the work of predecessors and assisted by various colleagues, Haldane began thinking of the body as a group of tissues, with each absorbing and releasing gases at different rates. Somewhat arbitrarily, he settled on five tissue groups, each with its own "half-time," or the period required for the group to become half-saturated—meaning, in a sense, half-full—at any particular depth. He used half-times of five, ten, twenty, forty, and seventy-five minutes. It was not that he believed that there were actual tissues with reliable half-times, but rather that an approach relying on imaginary tissues with specific characteristics could offer practical results. Tissue groups with half-times allowed Haldane and others, including Haldane's son, who undertook most of the mathematics, to calculate decompression requirements. The results of those calculations could then be tested on goats. Or other animals. Or people.

The experimental trials were clear. Goats and people could be decompressed reasonably safely by bringing them slowly to the surface. Because it was the relative change in pressure that mattered more than the depth itself, the ascent rate had to decrease as the diver came up. Haldane realized that this would be challenging in the field, out there in the water, off the side of a Royal Navy ship. Instead of slowly decreasing ascent rates with depth, Haldane turned to what became known as staged decompression. The diver would ascend at a steady rate, but would stop along the way. The first stop—at, say, forty feet—would be short, perhaps no more than a few minutes. The final stop, at ten feet, might drag on for an hour. The number and duration of stops would be different for different dives, for different combinations of depth and time, but the principle was always the same.

In preparation for actual ocean dives, goats were replaced by Lieutenant G.C.C. Damant and Gunner A. Y. Catto. Damant was the Royal Navy's inspector of diving, and Catto, in addition to being a gunner, was an experienced Royal Navy diver. The deepest of these chamber dives took the men to an equivalent pressure of 180 feet, where, after a descent lasting sixty-eight minutes—slow because of the project's limited air compressor—they stayed on the bottom for twelve minutes and then decompressed for fifty-one minutes. "In view of the results with goats," Haldane wrote of these chamber experiments, "the occurrence of decompression symptoms seemed probable." But the worst symptoms reported were itching of the skin, today often called skin bends. Neither Haldane nor the divers were concerned about itching, and, even now, skin bends, if not accompanied by other symptoms, are often considered benign.

In 1906, Haldane moved his experiments into the ocean, off Rothesay, Isle of Bute, working from the Royal Navy's 242-foot-long steam-powered gunship HMS *Spanker*. Neither Damant nor Catto had previously dived in actual water below 138 feet, an impressive depth for its time.

On the *Spanker,* air for the dives was provided by manually powered compressors, by sailors turning cranks on what Haldane described as "two double pumps," presumably meaning that each pump had two handles, one on each side. There were three men on each side, or six men per pump, all working so hard that they had to be relieved every five minutes. And even with that, the air delivered to the bottom was, at best, barely sufficient. Regarding one of Catto's dives, to 180 feet, Haldane wrote, "The rate of the pump could not be kept up above 24 revolutions per minute," and "he was almost overcome by the excess of CO_2."

There were other problems, too. On one dive, deep enough for significant nitrogen narcosis, Catto's hose became fouled. At 180 feet, where he was supposed to spend ten or fifteen minutes, almost twenty-nine minutes passed before he could clear his hose and ascend. Haldane extended the decompression, allowing ninety minutes from the time Catto left the bottom until the time he reached the surface. "There were no ill effects," Haldane wrote.

But in retrospect, after reviewing the experiments and realizing that the men dived day after day, building up levels of dissolved nitrogen each day, it seems unlikely that there were no effects at all. Haldane had already acknowledged that he did not take skin bends seriously, and in his written reports he dismissed symptoms of mild pain, noting that they generally cleared up on their own in both goats and men.

Before it was over, the divers reached 210 feet, a record for British helmet divers. Haldane mentions Greek sponge divers in the final paper summarizing his work, referring to those who were using helmets, but he does not mention the Royal Navy writings of Captain Spratt describing the deep breath-hold divers from

Greece capable of working below 210 feet on a single breath. If he had heard stories of these men, it is likely that he dismissed them, worthy of no more attention than tales of mermaids and sea monsters.

Haldane was not all work and no play. From the *Spanker,* under instruction from experienced divers, both he and his thirteen-year-old son descended to thirty-six feet. Why? To experience firsthand some of the realities confronted by divers and, probably, simply because they could.

In 1908—the year of the first Model T Ford, the reported slaying of Butch Cassidy and the Sundance Kid in Bolivia, and the world's first fatal airplane crash—Haldane and two of his colleagues published "The Prevention of Compressed-Air Illness." One of the other authors was Lieutenant Damant, one of the men who had descended to 210 feet. The paper, which included decompression tables and was to some degree a rehash of a report prepared for the Royal Navy, dismissed the slow but steady decompression recommended by others, asserting that staged decompression was both faster and safer. Divers and caisson workers could turn to the tables, look up a depth-and-time combination, and know how to limit the likelihood of bends. For example, a dive to 133 feet for twelve minutes would require decompression stops at 30, 20, and 10 feet, for a total decompression time of sixteen minutes. Decompression penalties climbed steeply with time spent at depth. A dive to the same 133-foot depth but with an added four minutes of time on the bottom would require four stops instead of three and a decompression time of thirty-two minutes.

The tables, extremely crude by modern standards, were immediately adopted by the Royal Navy and were used well into the 1950s. With some modifications, they are still used today.

Haldane's work was probably not known in the world of Greek sponge divers in 1913. But had Haldane been consulted about your imaginary helmet dive to 288 feet, he would have suggested no more than twelve minutes on the bottom followed by six decompression stops, starting at 60 feet, for a total decompression time of thirty-two minutes. He might have suggested an onboard pressure chamber, something to recompress you in the event of bends. And, undoubtedly, he would have wanted to be on-site to observe you. He would have wanted to hear how the dive went and how you felt afterward, and maybe to see firsthand whether you survived through the night.

It seems reasonable to believe that after Haldane's experiments, decompression was mostly worked out. Sure, there were blank spaces left to fill in, a matter of improving the mathematics, of improving the safety, but Haldane had made the final breakthrough.

While this may seem like a reasonable belief, it is not correct. In Florida, I attend a lecture by physiologist Virginie Papadopoulou, a young researcher from the University of North Carolina at Chapel Hill. She is a recreational diver—like millions of others, a person who dives while on vacation—but she is also interested in decompression theory. More than interested. She, like many other researchers scattered around the globe, conducts research on decompression. After her lecture, we talk.

"Decompression sickness is rare, and the guidelines work well," she tells me. "To a researcher, working on decompression sickness is like working on a rare disease. It is about understanding mechanisms."

She happens to have, on her computer, an ultrasound image of a bubble in some poor devil's heart. Even very tiny bubbles, she tells me, can be imaged, and in the future it may be possible to see the smallest of bubbles. But she cautions me against thinking too much about the bubbles themselves.

"Quick-onset bends might be caused directly by bubbles," she says. So, when a diver surfaces and feels immediate pain or collapses on deck, the diver is probably responding to a bubble blocking blood flow or putting pressure on a nerve. But divers who do not have immediate symptoms, those who do not feel pain until an hour or two hours or ten hours after a dive, may be responding not to the bubble but to the body's response to the bubble.

Bubbles, she explains, are flexible. They can and do move through blood vessels, changing shape as they travel. But they are also bumping into things as they move along. They are deforming tissues as they go. The body does not like this sort of thing. Bubbles bouncing around where they do not belong trigger an inflammation response.

Quick-onset bends can usually be treated successfully with immediate recompression. The diver goes back under pressure in a chamber, the pressure squashes the troublesome bubbles, the diver returns very slowly to the surface, and life goes on. In slower-onset bends, by the time a diver enters a chamber, the bubble or bubbles may already be gone. The chamber, along with the breathing of pure oxygen under pressure, treats the inflammation. Recovery may take repeated pressurizations and long periods breathing oxygen. Alternatively, recovery may just be a matter of time. The swelling may eventually subside on its own, and the victim, while not exactly none the worse for wear, might recover more or less completely.

There is still a need for ongoing trial-and-error research, for pressurizing divers and decompressing them at different rates to find the best approach. But the studies that most interest Dr. Papadopoulou are those that look at cause and effect, at underlying mechanisms. The future, she believes, may be in personalized decompression algorithms. In other words, instead of one mathematical model to fit all sizes and shapes of divers, algorithms might be adjusted for individuals. The beginnings of that future have already arrived. Divers routinely tailor their decompression schedules based on things like age, fitness, and water temperature. But those adjustments are seat-of-the-pants tinkering. What Papadopoulou sees as a brighter future would start with an understanding of what causes decompression sickness and move from there to customized decompression schedules and gases. But, she says, decompression research is like research on any rare disease, and funding for work on rare diseases is, well, rare.

"Our investigation," Haldane wrote, "which was planned with the more particular object of furnishing information required for securing the safety of divers ascending from deep water, was rendered possible by the gift to the Lister Institute by Dr Ludwig Mond, F.R.S. of a large experimental steel pressure chamber." The chamber that he used to experiment on goats and humans, including himself and his son, leaked, but his compressors—slow, inadequate, and inefficient by modern standards—kept up, and his experiments succeeded.

By 1908, pressure chambers large enough to host humans were an increasingly common sight not only among workers subjected to pressure in caissons and underwater but also in the arenas of medicine and medicinal quackery. They had, by then, been around for 246 years. The first is usually credited to a British clergyman named Henshaw. In 1662, he pumped air through organ bellows into what he called his "domicilium." Henshaw believed that increased air pressure would help many patients, offering a panacea of sorts for those with chronic lung disorders. Even the healthy would benefit from pressure. "In times of good health," he wrote, "this domicilium is proposed as a good expedient to help digestion, to promote insensible respiration, to facilitate breathing and expectoration and consequently, of excellent use for prevention of most affections of the lungs."

By the 1800s—a time when health crazes included water treatments at spas and the intentional drinking of crude oil— "pneumatic institutes" appeared in Europe. For a fee, patients could be compressed to pressures equal to those found between about thirty and one hundred feet of seawater. The compressed air, doctors told their patients, increased the circulation to the brain and internal organs. Such treatments were especially useful, proponents claimed, for disorders of the lungs and airways, diseases such as tuberculosis, emphysema, bronchitis, asthma, laryngitis, tracheitis, and pertussis. But these claims, all without basis, also extended to conjunctivitis, rickets, cholera, and deafness. A French surgeon built a pressurized operating room in 1879, reasoning that nitrogen narcosis could provide anesthesia. He was apparently unconcerned that surgeons, pressurized along with their patients, would be operating under the influence of nitrogen.

Americans, although slow to enter the game, were not to be outdone. In the 1920s, Dr. Orval Cunningham convinced himself that pressurization could help patients affected by the Spanish flu and those who had fallen into comas or whose skin was blue from their body's failure to deliver oxygen to tissues. He was not unduly discouraged when a mechanical failure brought one of his chambers unexpectedly to the surface, killing all of his patients. He continued offering treatments to those suffering from syphilis, high blood pressure, diabetes, and cancer. In 1928 in Cleveland, he built a chamber with a diameter of sixty-four feet. Its five floors each held twelve rooms that resembled those found in the better hotels of the era.

But these treatments had no basis in science. Europe's pneumatic institutes fell out of fashion. In Cleveland, Cunningham was asked for something resembling data that could justify his giant hyperbaric sanitarium. He provided nothing of value. "Under the circumstances," stated a report from the American Medical Association, "it is not to be wondered that the Medical Profession looks askance at the 'tank treatment' and intimates that it seems tinctured much more strongly with economics than with scientific medicine." In 1937, Cunningham's five-story chamber was torn down and the material sold as scrap.

Cunningham and his European predecessors might have been overzealous and uninterested in the niceties of science, and they might have seemed more interested in their patients' money than their health, but it turns out that their faith in what came to be known as hyperbaric medicine was not entirely misplaced. In 1955, Dr. Ian Churchill-Davidson administered oxygen under pressure, inside a chamber, to cancer patients. His goal was not to treat the cancer, but rather to treat certain side effects of radiation therapy. It worked. Within a few years, the use of oxygen in pressure chambers was shown to be an effective therapy for gas gangrene and carbon monoxide poisoning. It worked miracles on diabetics plagued with wounds that otherwise refused to heal. Within three decades, a committee of physicians provided guidelines for what became known as hyperbaric oxygen therapy, or HBOT.

In a response to the indiscriminate and overzealous use of hyperbaric oxygen therapy, the Undersea & Hyperbaric Medical Society, working with Medicare and private insurers, convened a committee and published a report in 1977 listing disease states that could and could not be appropriately treated with HBOT. Since then, this committee report, repeatedly updated, has become the standard reference on the conditions that are appropriate for HBOT—conditions such as aseptic bone necrosis, certain kinds of crush wounds, carbon monoxide poisoning, and, of course, bends, now most often called decompression sickness. But all of this was yet to come, and, with the exception of the use of chambers for treating diving accidents, it was all a side note to diving.

But there was no question that divers and caisson workers could benefit from going back under pressure when struck down by bends. By 1924, only sixteen years after Haldane published the decompression tables that were supposed to prevent bends but for various reasons did not always work as advertised, the U.S. Navy published treatment tables in its diving manual—tables that offered instructions on exactly how to repressurize bent divers and how to bring them slowly back to the surface.

Diving, in less than a century, had moved from almost complete reliance on individuals skilled at holding their breath, gifted in the art of apnea, to the routine use of pressurized air. Although divers breathing pressurized air could not go deeper than accomplished breath-hold divers, they could stay down longer and take on more work. But as they went deeper, their decompression obligations increased. Dives below two hundred feet were dangerous, logistically challenging, and short.

The kinds of dives needed for Operation Ivy Bells, or the kinds needed for offshore oil field construction, for work that required not minutes or hours but days or weeks on the bottom, required still more innovation, still more deaths, and still more breakthroughs.

Chapter 4

SATURATED

Lf the whole body of a diver were allowed to become saturated at any great depth," wrote John Scott Haldane in a magical sentence buried in his landmark paper of 1908, "it is evident that the time needed for safe decompression would be impracticably long."

So much of what Haldane said, wrote, and thought was right, but the words "impracticably long," it turns out, were off the mark. His idea of diving involved a copper helmet, a canvas suit, a long hose, and lots of lead. His experiments relied on air compressors powered by hand and a leaking chamber made from riveted steel plates. The technologies that made Operation Ivy Bells possible, that allowed divers to work in four hundred feet of water for days and weeks at a time, were, for Haldane, all but unimaginable. He had no way to foresee the revolution in equipment that would change everything, let alone the development of submarines that could stay submerged long enough to support an operation like Ivy Bells.

Ignore the second half of Haldane's sentence, the mention of "impracticably long." Instead, focus on the first half. Focus on the possibility of a diver's body becoming saturated at depth.

Any child who has dissolved sugar in water understands

saturation. Take a glass of water, add sugar, stir, and the sugar dissolves, disappearing into the water itself. Add more sugar, stir some more, and more sugar dissolves. The solution remains clear. But keep adding sugar and keep stirring, and eventually the sugar will no longer disappear. The recently added grains will whirl around in the glass. Stop stirring, and they will settle to the bottom. The water, saturated with dissolved sugar, can hold no more.

Lower a diver to, say, two hundred feet. Ignore, for the purposes of understanding saturation, the problems with breathing air at such depths.

Leave the diver at two hundred feet for thirty minutes, an hour, two hours, three hours. With additional time, more nitrogen dissolves in the diver's blood and tissues. The longer the diver stays down, the longer it will take to safely bring him or her back to the surface, to remove the body's load of dissolved nitrogen without forming bubbles and causing bends, to safely decompress. Leave the diver for four hours, five hours, six hours, and the decompression obligation continues to grow.

Leave the diver still longer. Leave the diver for, say, a day. Now the diver's body is saturated with nitrogen. For the diver to surface safely, to decompress, will require something like two days, two days during which he or she simply waits and breathes and releases nitrogen out of the blood and tissues while inching toward the surface. The diver, moving slowly upward, off-gasses.

But the beauty of saturation is that the diver need not surface after just one day. The diver can stay at two hundred feet for another day, another week, another month, another year, and then surface safely in just two days. Once the diver is saturated at a particular depth, the decompression obligation remains the same.

To make the stay at depth and the long decompression practical, different diving equipment and better chambers were needed.

Ideally, for example, saturation divers would want chambers that did not leak at the seams. But different equipment and better chambers would not be enough. An improved understanding of the gases that people could safely breathe was necessary. The nitrogen in air intoxicated divers at depth, and the oxygen, if breathed at extreme depths or for long durations, killed them. Air was just not the right stuff to breathe when diving deep and staying long.

Starting in the 1880s, an inventor and entrepreneur named Elihu Thomson, a man born in England but raised in Philadelphia, submitted one patent application after another. His U.S. Patent Number 261,790 described an electric arc lamp. His U.S. Patent Number 335,159 offered a means of distributing electricity. His U.S. Patent Number 403,707 covered electrical soldering.

Before he finished, Thomson held more than seven hundred patents, many but not all related to electrical equipment with substantial commercial value. He became wealthy. For a short time, he presided over the Massachusetts Institute of Technology as a somewhat reluctant president.

Whatever else he may have been, Elihu Thomson was a busy guy. But not so busy that he could not make time to suggest, in a letter to a friend, that divers might try replacing the ordinary air they breathed with a mixture of helium and oxygen. He wrote the letter in 1919, but he did not bother with a patent application.

That same year, another man, Charles John Cooke, ran with the notion that divers might be able to breathe something other than plain old air. He applied for what became U.S. Patent Number 1,473,337, which was ultimately issued on November 6, 1923. "The foremost object of my invention," he wrote in his patent application, "is to provide an improved atmospheric compound or breathing fluid to be supplied to divers while performing diving operations."

Pause here to consider the reality behind the suggestions coming from Thomson and Cooke. In 1915, almost a decade before Cooke's patent, U.S. Navy divers breathing air had reached 304 feet. They had gone to 304 feet to salvage a submarine, the USS *F-4*. And although they were there for work and not to set a record or to prove something to themselves and their friends, the depth they reached was a record that would stand for years to come. The U.S. Navy, in 1915, established a basement at 304 feet, a depth at which the divers were, at best, confused and easily distracted, and, at worst, hallucinatory, but still, at times, capable of meaningful work.

Thomson and Cooke lived in an era when diving meant bolting a person into a bulky suit weighing almost as much as the person himself. Diving beyond a few tens of feet meant using the recently publicized decompression tables of Haldane but still risking bends. Available air compressors struggled to provide the volumes and pressures required down there in the dark and the cold and the currents. It was against this backdrop that these two men, Thomson and Cooke, suggested breathing something other than air, something that sounded fine on paper but was entirely unnatural and untested.

The suggestion was not immediately adopted. Aside from the very outlandishness of their proposal, helium was, at the time, hard to come by. Out there in the universe it was the second most abundant element, but on Earth it was rare. And its rarity meant that, on a per-breath basis, it was simply too expensive to use.

But in time demand drove supply. The military wanted helium for lighter-than-air craft, for observation balloons and dirigibles. But where to find it? In wells, it turned out. Certain natural gas wells contained not only methane, propane, butane, carbon dioxide, and hydrogen sulfide, but also helium. It flowed from the ground in Kansas and Texas. Some wells produced so much helium and so little in the way of methane and its siblings that the raw product would not even burn. The price of helium dropped precipitously.

By 1924, the U.S. government was using animals to experiment with mixtures of helium and oxygen. That is to say, they put animals in pressure chambers where they breathed helium and oxygen instead of air. Initial results were discouraging. Cooke's improved breathing fluid left lots of experimental animals with symptoms of bends.

Researchers eventually realized that a light gas such as helium would behave differently in the body than a heavier gas such as nitrogen. It would diffuse through blood and tissues quicker. To compensate, initial decompression stops had to be deeper than those used on the same dives with the animals breathing air.

Among the researchers and experimental subjects were physician and diver Edgar End and his colleague Max Nohl, both civilians, both entrepreneurs, neither prone to irrational fears or, for that matter, perfectly rational fears. They undertook experiments on themselves in decompression chambers.

Based on chamber trials and theory, they decided to move to open water. On December 1, 1937, Nohl, encouraged and guided by End, descended into the cold depths of Lake Michigan, about twenty-five miles from Milwaukee. He broke with convention by breathing a mixture of helium and oxygen, but also by ignoring the tried-and-true deep sea diving helmet and suit in which divers had, by then, a century of experience. Instead, he used a suit and helmet of his own design.

Made from aluminum, his helmet looked something like the upper floor of a lighthouse, with windows in all directions, but with bars across the glass and a lifting eye at the top. There was no breastplate. The helmet attached directly to a tight-fitting rubberized canvas suit. The smaller volume of the helmet and suit in comparison with that of conventional equipment alleviated the need for a weight belt. To counteract the buoyancy of his gear, Nohl wore only a pair of detachable lead boots, each about eighteen pounds.

The suit looked nothing like conventional gear, but its differences extended well beyond outward appearances. Nohl was not attached to the surface by a hose. His breathing gas was stored in steel cylinders that he wore on his back. His link to the surface consisted of nothing more than a safety line and a telephone cable.

According to Dr. End, "In one of the steel cylinders is carried a respirable gas which the diver admits into the suit to equalize increasing water pressure as he descends. The other cylinder contains oxygen which enters the suit at a rate carefully controlled by the diver to satisfy his metabolic requirements."

Inside his aluminum helmet, Nohl breathed through a mask that fit around his nose and mouth, similar in appearance to those used by aviators. His exhalations moved through a canister of soda lime that chemically removed carbon dioxide. Nohl added oxygen to the breathing mix as necessary to sustain life. He was not only breathing the helium and oxygen mixture suggested by Thomson and Cooke, he was rebreathing it.

To any onlooker familiar with diving as it was normally done in the 1930s, Nohl must have looked like a suicidal crackpot. For that matter, to any onlooker familiar with diving as it is done today, Nohl would look like a suicidal crackpot.

Nohl submerged at 1:25 p.m. on that December day. At 1:34 p.m., he talked through his helmet intercom to the surface. "I've hit bottom," he said. His depth was 420 feet, more than 100 feet beyond the basement established in 1915, twenty-two years before.

Visibility was about six inches, but there was nothing to see. The lake bed was featureless clay. Nine minutes passed, and then he was on the way up. His ascent required 118 minutes.

On the surface, tenders removed his helmet. Nohl, according to newspapers, asked about what he perceived as a strange smell. "That's fresh air," his tenders told him as reporters snapped photographs.

On the bottom, he had not experienced the tunnel vision or convulsions of oxygen toxicity. Nor had he experienced even a hint of the narcosis that made men breathing air see railroad tracks and trains and mermaids and monsters at much shallower depths. On the bottom, the water had been cloudy, but his head had been clear. And upon surfacing, Nohl experienced no pain, no numbness, no paralysis. And he did not die.

The depth barrier created by breathing air had been smashed. Human beings breathing pressurized gases were no longer trapped in the shallows. And at the same time, on the same dive, Nohl had reminded divers that the heavy helmets and suits they had been using might not be the only gear for the job. There could be other ways to dive.

⌐

Nohl was not the only diver to develop a new kind of diving helmet. Anyone clever with tools could modify or build diving equipment, which, after all, was often little more than an inverted bucket secured to a suit. Some divers were going without the suit, diving by simply keeping their head inside an upright helmet. Even kids were building their own. "They go down to the sea in old water heaters along the Atlantic coast these days," a 1932 article in *Modern Mechanics* said, "now that some young man with a leaning toward aquatic sports has proved how easy it is to make an excellent diving helmet from a metal water heater which will enable its wearer to walk comfortably on the sea floor 35 feet and more below the surface." The article explained how to make not only the helmet but also the compressor, in this case just a manually operated

116

bellows of wood, leather, and tin that could be used to pump air through a garden hose. Divers would have to work in pairs, one underwater and the other on the surface, pumping.

A Diving Helmet
from a
Water Heater

Air for the diver is supplied from twin bellows.

THEY go down to the sea in old water heaters along the Atlantic coast these days, now that some young man with a leaning toward aquatic sports has proved how easy it is to make an excellent diving helmet from a metal water heater which will enable its wearer to walk comfortably on the sea floor 35 feet and more below the surface. A few feet of garden hose, two pairs of bellows, a couple of valve boxes and a cylindrical metal boiler of the type used in most homes for heating water, are the essentials for building one of these helmets.

Completed water-heater diving helmet.

boards drill a pair of one-inch holes to admit air. On the inside of the board, covering the holes, tack a piece of chamois by its four corners to act as a valve.

Gores for the bellows are cut out of leather in accordance with the printed pattern. Moderately thin and very flexible leather should be

Over the past century, thousands of different kinds of diving helmets have been made and used. Many were manufactured to specific design standards by qualified craftsmen, but others were fashioned from old water heaters, as described in this 1932 Modern Mechanics *article. (Image from* Modern Mechanics *magazine courtesy of John Schroeter)*

The article did not say so, but if the diver working the bellows grew impatient, he had only to stop or slow his pumping, and his turn would come quickly.

The innovations were seemingly endless. Many diving suits and helmets were designed but never built. Some were built but never used for diving. Some were built but used only once, with fatal results. Many others, however, were built and used repeatedly. Most, but not all, relied on compressed air. Most, but not all, relied on a hose coming down from the surface, a hose through which pressurized air could be pumped to the diver.

There was also scuba—self-contained underwater breathing apparatus—which did not require a hose to the surface. Scuba divers were fully exposed to pressure, but the hoses, like the one used by Max Nohl, ran only a short distance, from a tank or tanks usually worn on the back to the diver's mouth and sometimes his or her nose. Its invention is often attributed to Jacques Cousteau and Émile Gagnan in 1942, but this attribution exaggerates the facts. Cousteau and Gagnan were innovators, strong contributors to the evolution of the gear, but their invention had many predecessors, including the Rouquayrol-Denayrouze system, which had been mass-produced as early as 1865. What made Cousteau and Gagnan's innovation unique was that it caught on. With very little training, divers could strap on a steel tank holding pressurized air and breathe through a regulator that opened and closed with each breath. Hundreds, then thousands, then millions of people took to the water. More relevant to the thread of the story of Operation Ivy Bells and the evolution of saturation diving, the U.S. Navy and others put scuba to routine use, making it another tool in an ever-growing toolbox. And the scuba of Cousteau and Gagnan was used, along with other gear, by the Navy's Sealab program, the first systematic attempt at saturation diving, the program that made Ivy Bells possible.

The Rouquayrol-Denayrouze self-contained underwater breathing apparatus, an early version of scuba used in the middle of the nineteenth century, as illustrated in the Swedish Nordisk familjebok *(Nordic Family Book), an encyclopedia of sorts, in 1876. The diver's self-contained air supply was supplemented by a hose running to the surface, but the hose could be and sometimes was removed. Air tanks of the time held very little air, so dive times were extremely limited when the hose providing air from the surface was removed.*

Enter Dr. George Bond, a young Navy diver named Bob Barth, and more goats. There were others—many others—involved with what began as Project Genesis and evolved into the Sealab program, but Bond, Barth, and the goats provide a starting point.

In 1957, Bond was a forty-two-year-old Navy physician accustomed to working with divers. He was known for quoting the Bible, and he sometimes read prayers to men inside decompression chambers, but by his own account he also drank and swore. He was not just a doctor but also a trained diver. When

Navy authorities did not prevent it, he was happy to experiment on himself, in the tradition of Haldane and Nohl. For example, in 1959, to prove that crews could escape from disabled submarines even at significant depths, Bond and a colleague swam through the escape trunk of the USS *Archerfish* to ascend 302 feet in fifty-two seconds. This involved flooding the escape trunk with seawater to pressurize the air inside, which allowed the men to open the hatch and begin their ascent, exhaling forcefully on the way up to avoid death as air expanded in their lungs. Despite the facts that this is theoretically possible and that Bond and his colleague succeeded, the mere thought of it should and does invoke chills in anyone familiar with the dangers of so-called free ascents. Swimming upward with a chestful of compressed air can cause the lungs to overexpand, to burst, sending a bubble directly into an artery and toward the brain. But Bond and his colleague exhaled forcefully enough, continuously, to avoid overexpansion of the lungs and to surface uninjured. For their trouble, they were awarded the Legion of Merit.

Bond believed, or said he believed, that the survival of humanity depended on an ability to colonize the seas, to grow food underwater. "Those of you who are close to the agricultural picture know," he once told a group of honors students at Albany Medical College, "that with the expanding populations which we have, should we avoid all-out war, and God willing we will, we are going to come up against the old enemy hunger, and it's gonna be serious. There's no reason to believe that we can take care of the combined hungers of the expanding populations a hundred years hence. The oceans around us can do it, however."

He liked to tell people that an ability to colonize the continental shelves would unlock an unexplored and unexploited region as big as Africa. Saturation diving—if it worked, if one could overcome the challenges of breathing gases and developing safe decompression schedules and simply keeping an underwater colony running for a reasonable amount of time—would open up an underwater continent.

The idea had, in fact, been tried at least once. Max Nohl—the same man who made a record-setting dive to 420 feet in an experimental diving suit in 1937—had already done a saturation dive, spending twenty-seven hours in a chamber breathing air at a pressure equivalent to 101 feet of seawater. But Nohl's team had concluded, according to many sources, that longer-duration saturation dives would not be possible.

In Bond's early Project Genesis research, in 1957, he used a chamber to expose experimental subjects to the pressure found in 198 feet of seawater. Plenty of people had been to 198 feet and deeper, both in chambers and in the open ocean, but none had stayed very long. Decompression obligations limited their bottom times. For dives at that depth, decompression times were usually far longer than bottom times. But Bond took his subjects to 198 feet and left them there for thirty-five hours.

The experiment offered both bad and good news. All of the experimental subjects died, their lungs covered with lesions. But on the upside, the subjects were rats—male albino Wistar strain rats—a variety of the plain old brown rat, *Rattus norvegicus,* bred for laboratories, with a pedigree dating back to 1906.

Bond tried again, with new rats. This time they breathed a mixture of just three percent oxygen and ninety-seven percent nitrogen. A mixture of three percent oxygen will not support consciousness on the surface, but at depth, where the pressure of the oxygen is elevated even though its concentration is low, three percent is more than adequate. This time, the subjects stayed at depth for fourteen days. Fifteen out of sixteen rats survived.

Bond and his furry friends tried different gas mixes and

different depths. The work was hard on the rats, but Bond learned that oxygen levels would have to be low in saturation. Otherwise, lesions formed on lungs. What Bond was seeing was a second form of oxygen toxicity. In acute oxygen toxicity, divers exposed to high pressures of oxygen experience sudden convulsions. They are overcome by tonic-clonic seizures. At lower but still elevated pressures of oxygen, breathed for long periods, divers may not experience convulsions, but instead their lungs might be damaged by what is sometimes called chronic or pulmonary oxygen toxicity. In essence, the oxygen burns the lungs. Mild chronic oxygen toxicity causes swelling in the airways and renders breathing difficult. It leaves divers feeling asthmatic. Deep breaths become unbearably painful. More severe oxygen toxicity slowly kills divers, whether rats or people.

High levels of nitrogen might be okay for rats, but the nitrogen would cause narcosis in humans. So Bond mixed low concentrations of oxygen with helium. Saturation divers would breathe not air, and not a mix of oxygen and nitrogen, but a mix of oxygen and helium, a mix that became known as heliox.

Bond soon graduated to guinea pigs and spider monkeys. Why spider monkeys? "In anticipation of future experimental work in which human subjects might be utilized," Bond later wrote, "it was felt desirable that small primates be exposed to the experimental conditions previously described."

And then there were the goats. In 1962, Bond showed up at the New London, Connecticut, submarine base with a small herd of goats. They were easier to manage and less expensive than spider monkeys, and closer in size to humans. And they had been used by Haldane. What was good enough for Haldane was good enough for Bond.

"Our earlier experiments were concerned with the feasibility of using an artificial atmosphere of oxygen and helium under pressures as great as 200 feet of water," Bond wrote. "A further problem was to determine decompression requirements in a minimal number of stages after prolonged exposure to severe atmospheric pressures." In other words, the first experiments showed that small animals—and, therefore, presumably humans—could breathe heliox for long periods at pressures equivalent to almost two hundred feet of seawater, but more experiments were needed to figure out how to bring a mammal the size of a human back to the surface without decompression sickness, without bends.

When Bond and his goats showed up at the submarine base, they met Bob Barth, one of several divers tasked at the time with training crews to escape from sunken submarines. Bond needed experienced men to help with the chamber experiments.

"During this period," Barth later wrote in his delightful memoir, *Sea Dwellers,* "someone was needed to tend those goats and to remove the goat pellets and other tidbits that you might find in a chamber full of live goats."

The experiments proved successful, or at least successful enough. In 1962, the Navy and Bond decided that it was time to put human volunteers into saturation. The plan was, in essence, to go slow. The first human volunteers were placed in a chamber at one atmosphere—at the surface—to test the effects of breathing heliox for long periods. Helium had its issues. It robbed heat from the body and distorted speech, making the saltiest divers sound something like Donald Duck. And no one knew what would happen to divers breathing heliox for long periods, for, as it turned out, 144 hours, or six full days and nights.

Two of the volunteers were doctors. The third was Bob Barth. "The third individual," Barth wrote of himself in *Sea Dwellers,* "didn't need a medical background but it might be important that he knew a bit about chambers. It also might help if the third individual wasn't the brightest sonofabitch on earth but seemed willing to undergo physical abuse." For six days Barth taught his two roommates to swear like proper sailors. He also underwent physiological and psychological tests. "If you weren't being asked to exhale everything you owned into a spirometer, you were reading ink blots or maybe giving a gallon or two of blood (a few CC's at a time) to vampires lurking outside."

In 1963, in what was known as Phase D of Project Genesis, Bond and his associates moved their work to Panama City, Florida. There, at the U.S. Navy Experimental Diving Unit, they had access to a more comfortable chamber. This is not to say it was a comfortable chamber. It did not have enough bunks for all the divers, so they would have to sleep where and when they could. Barth and three companions were pressurized to the equivalent of one hundred feet of seawater for six days and were breathing seven percent oxygen, seven percent nitrogen, and eighty-six percent helium.

The physiological and psychological tests continued. "I never saw the urge by so many to obtain the blood of so few," wrote Barth.

The next experiment took the divers to a pressure equivalent of 198 feet of seawater, the same pressure that had killed every one of Bond's first batch of rats, the experimental subjects that had breathed pressurized air. But the human divers, like the later groups of rats and guinea pigs and spider monkeys and goats, were breathing heliox, what Bond sometimes called "a synthetic atmosphere." The experiment went off without a hitch.

"As a result of some six years of animal and human studies involving closed ecological systems, elevated pressures, and synthetic atmospheres," Bond wrote, "the stage has been set for operational application of the work. It would now appear that we can safely station men at any point on the submerged continental shelf, with a reasonable expectancy of useful performance for prolonged periods of time."

Project Genesis ended, and the Sealab program began. Their next saturation dive would be in actual water.

Bond's systematic approach, his ever-growing collection of test results, set his work apart from that of other groups pursuing saturation diving. There were, for example, the saturation dives supported by Jacques Cousteau. In 1962, Cousteau and his team saturated two men in a habitat they called Conshelf I off the coast of Marseilles. For one week, the divers were saturated a mere 33 feet from the surface, where they spent their days looking at fish, symbolically cultivating a subsea garden, and smiling for the cameras. In 1963, Cousteau's team saturated more divers, this time at 33 feet for a month and at 100 feet for a week. In 1965, Cousteau's team saturated six divers, this time at 330 feet.

As another example, there were the saturation dives supported by inventor, entrepreneur, and underwater explorer Edwin Link. In 1964, Link's team put two men into saturation for forty-nine hours at 432 feet. Their habitat was little more than an inflated bag pinned to the seafloor. They called it SPID, for Submersible, Portable, Inflatable Dwelling.

These early saturation dives went forward without serious mishaps.

In contrast, Swiss physicist Hannes Keller and British journalist Peter Small were not interested in saturation but rather in reaching extreme depths using a variety of different gases intended to accelerate decompression. On December 3, 1962, Keller and Small climbed into a diving bell, closed the hatch, pressurized their atmosphere, and were lowered to one thousand feet. At that depth, Keller opened the hatch and left the bell, planting flags on the continental shelf. He became entangled with his flags and lost precious time, but he managed to disentangle himself and return to the bell. With Small's help he closed the hatch.

In a hurry and possibly already realizing that things were not going as planned, Keller opened a valve. The valve he should have opened would have provided a specially engineered gas mix. Instead, the one he did open provided air. Both Keller and Small lost consciousness within minutes. Their surface support crew, watching all of this through cameras, raised the diving bell. When it reached two hundred feet, the men inside were still unconscious, but it became clear to those in the support ship that the bell was losing pressure. Two surface divers, one of them only nineteen years old, went down to solve the problem. They realized that a tip of one of Keller's fins was stuck in the hatch. They cut away the fin tip and managed to seal the hatch. But in the process, one of the surface divers, the nineteen-year-old, disappeared. His body was never recovered.

The bell, with its two unconscious divers, was raised to the surface. Both men eventually regained consciousness, but for Small it was a temporary victory. He died before completing his decompression. His wife, distraught, took her own life nine weeks later.

Bond knew and worked with, to one degree or another, Cousteau and Link and Keller. He considered Cousteau to be among the most charismatic men he had ever met. He offered all of them warnings about moving too quickly, about going from animal experiments to human trials without due process. He and his team were, he wrote, "determined to conduct our work in an orderly, carefully documented, scientific manner that would progress to our goal of placing men and women on the ocean floor to live and work freely." His primary interest was not in being the first or the deepest, but in being the one with the greatest understanding of the realities behind what was a breakthrough in human capabilities. He remained, above all else, devoted to the painstakingly methodical and therefore slow investigation of the physiological effects of deep diving.

Bear in mind that the Sealab program was progressing in the early 1960s, when, among other distractions, NASA was supplying America with heroes. In 1961, Alan Shepard became the first American in space, followed by Gus Grissom, and then, in 1962, by John Glenn. The public was enthralled. These astronauts were not only explorers, they were in an exploratory race, at first behind but then neck and neck with the Russians. This was front-page stuff, headline news. Their magical visuals included homeland Earth, a planet small and round and insignificant, while also huge and blue and all-encompassing. The U.S. government wanted public idols and got them. Parades featured the guys with what writer Tom Wolfe would later famously call "the right stuff," writ large. These were young Americans willing to sit on top of experimental rockets that hurled them into the vacuum of space, where they circled around the earth before risking fiery-hot, suspense-ridden reentries. Never mind that monkeys could do and had done the jobs of early astronauts, these men were credentialed, recognized by the U.S. government at its highest levels.

To the extent that there was room in the spotlight for saturation diving, for exploring and living in the sea, it went to Cousteau, aided by American journalist James Dugan. But, at best, even Cousteau and his teams stood at the shadowy edge of the light. They offered visuals of fish and spacecraft-looking diving systems, but the deep sea was not space. These were not landscape shots showing the planet on which all of us lived, but rather close-ups and portraits against backgrounds of sand and coral. Even with Cousteau's charming French accent, the underwater world could not ignite the fervor touched off by an Atlas rocket.

Sealab struggled. Bond's bosses misunderstood the value of what he was doing. One of his proposals was rejected with a note saying that it "was ill-conceived and impossible to accomplish" and that his time would be better spent working "on matters of operational importance." Bond and his divers might have had the right stuff, but they were using it in the wrong way. They were going in the wrong direction. They were swimming against the tide of public exuberance.

At times, Bond and his colleagues had to open their own wallets to pay for basic supplies. And when it came to building the underwater habitat that would become a temporary home for Bob Barth and three other divers, they were forced to go around ordinary Navy procurement procedures. The hose that would provide Sealab I with fresh water, for example, came from the gardening department at Sears. For other parts, they turned to a scrap yard, undertaking what Barth later called "midnight requisitioning."

One person struck by Sealab's tight budget was astronaut Scott Carpenter, who joined the Sealab program in 1964, during the preparation of Sealab I. He would, in a sense, be a link between it and NASA, between inner and outer space. He would also be the program's celebrity diver.

From Bond himself: "As investigators, we weren't given priority for material or equipment; we took what we needed from any place that had it." And this: "It was rumored that base personnel locked their doors when they saw us coming."

They were not, to be clear, stealing. They were just moving materials from one Navy operation to another, and all for a good purpose.

The hull for Sealab I was built from abandoned cylindrical floats, each fifty-seven feet long, originally intended to support minesweeping operations. Bond and Barth found the floats rusting in the Navy's Panama City salvage yard.

The floats that would become Sealab I looked something like gigantic propane bottles. In principle, the design was simple. Cut one end off each cylinder and weld the cylinders together to make a single long enclosed tube. Then cut a hole in the bottom. Add legs so that it could stand on the seabed. Install bunks and a simple galley and a toilet. Put in windows and odds and ends of plumbing, some of it purchased at retail outlets, where it could be had at bargain-basement prices. And remember to attach weights so that, when full of helium and oxygen, it would sink. With such a tight budget, the weights used included scrapped train axles.

The first descent, off the coast of Panama City in a mere sixty feet of water, was a trial. According to Barth's memoir, the rope that was meant to control the habitat's descent stretched. The habitat sank uncontrollably and flooded. This was not quite as bad, as, say, a rocket blowing up on the launchpad. No one was hurt. Divers raised the habitat, towed it back to a dock, dried it out, and replaced anything that had been damaged beyond repair.

The second attempt to lower the habitat off the Florida coast was successful. Sealab I passed its initial in-water test. It was raised, loaded onto a barge, and shipped to Bermuda.

Inspired by Cousteau and the French approach to life underwater, the habitat was stocked with, among other things, wine. But the French approach was not replicated completely. The divers would subsist on canned food and, unlike Cousteau's team, would not have a chef with them in the habitat.

On July 20, 1964, Sealab I was lowered to a depth of 193 feet next to a Navy platform near Bermuda. When the lowering and setup were complete, Bond himself dived for a final inspection. The water was so clear that he could see the entire habitat on the seabed. He surfaced smiling.

The saturation divers, including Bob Barth, climbed into a diving bell and were lowered to within a few feet of the bottom. Sealab I, their home for the coming days, stood seventy-five feet away. The divers chose to swim to the habitat, each carrying a small bag of personal items, on a breath of air.

"On the afternoon of 20 July 1964," wrote Barth in his memoir, "George Bond's dream of men living on the bottom of the sea became a reality." The dive was scheduled for three weeks. The oddity of it all occurred to Barth, and probably to his three companions. Here they were, almost two hundred feet down, with reasonably comfortable living quarters and food and wine. Occasionally, Bond, affectionately dubbed Papa Topside by his men, read poetry or quoted scripture over Sealab's speakers. All of the divers read. They talked to ham radio operators via a link to a shortwave radio on the surface. Here they were at 193 feet underwater, and no one was watching the minutes tick by, no one was worried about racking up an impossibly long decompression obligation.

As they had done during chamber dives, the divers tolerated repeated physiological tests and provided seemingly endless samples of various bodily fluids. But when there was time—and they found the time—they ducked outside. Sometimes they swam around on a breath of air. Sometimes they used ordinary scuba gear. Sometimes they used a Navy rebreather, equipment similar in principle to that used by Max Nohl in his 1937 dive to 420 feet, though their version looked more like scuba gear and did not require the lighthouse-looking helmet Nohl had used.

Outside, the divers swam around, unhurried. They felt and acted as if they belonged there. They opened cans of sardines and fed them, sometimes one at a time, to amberjacks and groupers. "The sea," wrote Barth, "was ours."

"The men, during the occupancy of Sealab, accentuated their personal idiosyncrasies," wrote Bond in a report. "During one period, excessive use of foul language developed, as well as an independent attitude with respect to the surface support." But, Bond noted, their apparent cockiness and swearing did not cause any safety concerns. The men below functioned as a team.

Bond's report does not mention the wine reported by Barth, which could explain the "personal idiosyncrasies" and swearing.

The operation was not without incident. One of the divers, Tiger Manning, was using a rebreather when he lost consciousness just beneath the habitat. He was hauled inside and revived.

And the entire operation was cut short by a brewing storm. After eleven days at depth, the divers decompressed for 31 hours and 35 minutes. That is, they sat in a chamber, breathing as the pressure was slowly decreased, as they were inched back to the surface, off-gassing.

Once they were outside, reporters greeted them. Barth, in his memoir, recalled wanting nothing more than to find a toilet, but he was instead forced to participate in a press conference.

A few days later, the *New York Times* ran sixteen paragraphs under the headline "Navy Men Set Up 'House' Under Atlantic and Find Biggest Problem Is Communication." The newspaper was as fascinated by the speech distortion that comes with helium as it was with the accomplishment of living at depth. Sealab I was finally attracting media attention, but the media did not convey the magnitude of the project's technical and scientific importance. There was no sense of what it meant for the future.

After Sealab I, Bond had little hope for further progress. His pessimism was not driven by technical or physiological limitations, but by the challenges of funding.

"My assessment of the situation was realistic," he wrote. "We had accomplished an exceptional exercise but we would not have the financing to extend our diving and working depth to the average depth of the continental shelf—about 600 feet." He also expressed this opinion: "It is because of the ultraconservative, nihilistic attitude of many people in the Navy that diving practices and capabilities made no appreciable progress during the past quarter century."

But Bond was surprised when he was called to a meeting with an admiral. The work would go forward after all. Why? Because in 1963, the USS *Thresher* sank, making her not only the first nuclear-powered submarine to be lost at sea but also one of only two submarines to kill, in a single accident, more than one hundred people. Also disconcerting was the fact that she sank while undertaking depth trials, more or less in communication with nearby surface vessels, and not while in combat or even on patrol. In the aftermath of the sinking, the Navy realized that it needed to up its game underwater. The fact that the *Thresher* sank in depths well beyond those that could be reached by divers did not stop the Navy from suddenly recognizing the value of the Sealab program and the potential of saturation diving.

Bond's team moved quickly. In 1965, a mere year after Sealab I, some four hundred people were involved in getting three teams of ten divers each into saturation. The cramped bargain-basement conditions of Sealab I gave way to Sealab II's roomy eating and living quarters, a dedicated laboratory space, and curtains strung across the habitat's eleven portholes. But whatever comfort was gained inside was lost outside. Unlike Sealab I, which sat in the clear, warm waters of Bermuda, Sealab II sat in 205 feet of water on the edge of Scripps Canyon in the cold, dark waters off La Jolla, California.

Among those inside was former astronaut Scott Carpenter. In what amounted to little more than a publicity stunt, a radio link was set up to allow Carpenter to talk to the crew of Gemini V, which was, at the time, orbiting the earth. Carpenter, breathing helium, sounded like a cross between a duck and a chipmunk.

Later, with a telephone link and amid considerable confusion with operators, Carpenter, in the same duck-chipmunk voice, chatted with President Lyndon Johnson. "I must apologize," Carpenter said, "for the sound of my voice, but it's the absolute best I can do in a helium atmosphere."

Johnson's responses to Carpenter, which amounted to various iterations of "We're mighty proud of you," gave no indication that he understood a single word.

Carpenter and the other divers provided blood, urine, and

breathing samples. Working outside the habitat, they were sometimes assisted by a trained dolphin named Tuffy.

Sealab II was not without problems. Carpenter was stung by a scorpion fish. Several divers, for various reasons, lost control of their buoyancy and barely avoided unintentional ascents that would have almost certainly resulted in fatal decompression sickness. Bond was aware of dangerous lapses in protocols that he attributed to "aquanaut breakaway phenomenon," or the saturation divers' tendency toward arrogance. And several divers reported symptoms of decompression sickness during their final slow ascent to the surface.

Later, Sealab II researchers wrote of the project, "A major purpose was to evaluate the diver's capability for doing useful work in the open sea, leaving from and returning to the habitat." And they observed, "The data analyzing the personal and social attributes of the divers who achieved good work records in the water suggests that divers should be of a relatively phlegmatic disposition and able to contend with stress and danger without being preoccupied with them."

By and large, Sealab II was a resounding success.

Four years stretched between Sealab II and Sealab III. The Sealab II habitat was modified to become Sealab III while test dives were conducted in chambers ashore. And then, on February 15, 1969, the new habitat was lowered to a depth of 610 feet near San Clemente Island, California. It was the same year American astronauts would take their first steps on the moon.

Sealab III was pressurized using heliox with its hatches sealed and no divers aboard. If all went as planned, divers would enter their temporary home on the bottom. But within hours, it was clear that everything was not going as planned. Sealab III was losing gas.

On February 16, Bob Barth, Berry Cannon, Richard Blackburn, and John Reaves were lowered to Sealab III in a diving bell. Barth and Cannon left the bell, only to discover that Sealab III's entrance hatch was jammed in the closed position. In a second dive, Barth and Cannon were to try the hatch again. Barth remembers leaving the bell and pausing for a second next to Cannon, admiring the view, seeing not just a landscape but a dream come true.

The two men moved toward the habitat. While Barth tried to pry the hatch open, Cannon, nearby, lost consciousness. A moment later, Bond, watching the dive on a television screen, saw

Sealab III, hoisted out of the water. (NH Series, Other, K-8149, Archives Branch, Naval History and Heritage Command, Washington, DC)

Barth dragging Cannon back to the bell. Reaves and Blackburn, in the bell, gave Cannon mouth-to-mouth resuscitation and CPR. Fourteen minutes later, the men in the diving bell called the surface. "Berry Cannon," they said, "is dead."

The men were decompressed. Sealab III ended before it really began.

Cannon and Barth, working at six hundred feet, had been using rebreathers, the gear that passed exhalations through a scrubber intended to remove carbon dioxide, injected oxygen to replace that used in metabolism, and sent the gas back to the diver. But it appeared that Cannon's scrubber canister, which should have held a chemical carbon dioxide absorbent, was empty. He was breathing his own exhausted carbon dioxide. At that depth, Bond later wrote, a diver breathing his own exhalations would lose consciousness within minutes. That is exactly what happened to Cannon.

Cannon's death spelled the end of the Sealab program. The Navy could not accept the risk of further dives. Or so it was said at the time. But that was simply not true. The Navy was happy to take the risk, though only in secret and with a military purpose. The Sealab trials prepared the Navy for Operation Ivy Bells. In other words, saturation diving did not die with Berry Cannon. It merely went underground.

The Navy put saturation systems on submarines, penetrated the territorial seas of adversaries, and, among other things, tapped their communication cables. Saturation diving had gone from a wild dream, dismissed as "ill-conceived and impossible to accomplish," to something of operational importance. Bond had seen saturation diving as a way to colonize the continental shelves, but the Navy had other ideas. They had Ivy Bells.

But the story does not end with Ivy Bells, which was in most ways little more than an interesting sidebar, a James Bondish use of what George Bond and his colleagues had developed. The Navy moved away from habitats and instead kept the pressurized living chamber—where saturated divers slept, ate, showered, and waited around—strapped to a submarine or on the deck of a surface vessel. Divers living under pressure in chambers on surface vessels would transfer to underwater work sites via a pressurized diving bell. The bell would be lowered to the dive site and the divers, once at a depth with pressure equal to that found inside their bell, could open a hatch and move outside. Habitats, for most purposes, were inefficient, expensive, and inflexible.

Bond, after the Sealab III fatality, was assigned to Panama City. There he developed the Ocean Simulation Facility, a complex of pressure chambers where divers could be saturated without the complexities of the open ocean. His work was once again confined to chambers. With Sealab III scrapped, Bond continued to experiment, both at the facility that he helped develop and at sea, but he would never again work on something as ambitious as Sealab, and by the time he left the Navy, he was, based on his memoirs, somewhat embittered.

George Bond died thirteen years later, in 1983, but the facility in Panama City lives on. The U.S. Navy Experimental Diving Unit is a place where military divers go to imitate Wistar strain rats and goats.

I arrange with the Navy for a tour and see pressure chambers big enough to house dozens of men at a time, some that can be flooded—so-called wet pots, where divers can swim, test gear, and pretend to be in the ocean without the inconvenience and expense of actually being in the ocean. The divers are subjected to low and high temperatures, and different breathing gases and decompression rates, while using various kinds of equipment. Their bodies are instrumented, and their bodily fluids are frequently sampled. They are rendered extremely uncomfortable and sometimes hurt. It is not especially unusual for divers at the Experimental Diving Unit, or EDU, to suffer decompression sickness, followed immediately by recompression and treatment.

Most but not all of the scientists are civilians. At times civilian divers work with the EDU, usually during equipment evaluations, but all of the divers I meet are military personnel who have served other assignments as divers before joining the EDU. Once at the EDU, they have the opportunity to volunteer as subjects in experiments. No one forces them to participate, and even after an experiment begins, they can pull out at any time, for any reason, or for no reason at all. They are not, after all, Wistar strain lab rats.

I spend the day hearing about careers, about kids from places like Kansas and Colorado who had turned eighteen before setting eyes on an ocean, but who, nevertheless, joined the Navy and became divers. Salvage jobs are a frequent topic of conversation. I am told, too, about the testing of diving equipment, some of which had been previously tested but in the field was involved in a fatality.

The accidents, a researcher tells me, are especially tragic "because they were so avoidable." One example: a diver surfaced and passed his breathing gear to a tender on the support vessel, and then promptly sank to the bottom and drowned. Another example: a diver working under a ship had trouble with his gear, so he surfaced on the far side of the ship and, floating, removed his helmet. His tender, who could no longer see or hear the diver, did what he was trained to do. He pulled the diver in. In this case, that meant pulling the diver, who was no longer wearing his helmet, under the ship. The diver drowned while being dragged back to the dive station.

In a day of interesting discussions, two conversations stand out. The first is with a senior diver, a man close to forty years old. After completing the Navy's diver training, he was deployed to Hawaii, then to the Persian Gulf, and then back to Hawaii. In 1997, the Navy trained him as a saturation diver. He made dives to one thousand feet both in chambers and in the ocean. He worked from submarines "doing stuff," as he puts it.

"Stuff like tapping communication cables?" I ask.

I get the answer I expected. "Just stuff," he tells me.

"I've heard," I say, "that the Navy has permanently installed listening stations all over the world—you know, SOSUS stations. Did you do any of that work?"

He cannot say. Or if he did say, he would have to kill me. He can tell me that, in all, he made twelve saturation dives. Since saturation dives typically run about thirty days, this likely means that he has spent about one year of his life at depth, saturated. He can also tell me that, below three hundred feet, his thumbs lock up, apparently due to fluid being squeezed from between his joints. And he can tell me about the cluster headaches that hospitalized him after one dive.

From George Bond: "If it were not for the development of saturation diving, the offshore petroleum industry as we know it would not have been feasible." Saturation diving, he wrote, "opened up the oilfields of the continental shelf for exploration and exploitation."

I talk with the senior Navy diver about my own experiences with saturation diving in oil fields. We trade notes about injuries, the tedium of multiday decompressions, the peeling of skin from hands and feet caused by constant dampness, fungal infections, the feeling of general malaise after surfacing that is usually attributed to the change from the synthetic atmosphere of the chamber to the real atmosphere of the surface. We talk of the bone-chilling cold that comes with breathing helium mixes, a hypothermia that seems to start on the inside of the body and work its way out. I tell him of dreams I had in saturation, dreams in which I pried open a chamber hatch and went on deck to sit in the sun, all the while hiding from my supervisor.

This is the kind of conversation between divers that could, especially in the presence of beer, go on for hours, but here we are on a schedule. Time is limited.

I ask if Navy divers ever grow frustrated with the knowledge of the very high wages earned by oil field divers, who are, in practical terms, doing the same work that he and his colleagues do for Navy pay. No, he tells me. No one he works with is in it for the money. But he is frustrated by the Navy's current lack of interest in saturation diving. At the time of our conversation, the Navy does not have an operational saturation diving system. "You look at other countries," he says, "and see what they're doing and wonder what our Navy is thinking." The South Korean navy, he tells me by way of an example, has a saturation diving program. "If we need to work at saturation depths," he says, "we will have to contract civilian divers." The disgust in his voice regarding the possibility of using civilian divers is as tangible as cold water. But he thinks that the Navy will see the light and reactivate a saturation diving system in the near future.

The second conversation that stands out is with Dr. John Clarke. He is not just a scientist but the EDU's scientific director, a civilian who has worked with the Navy since 1979 and with the EDU since 1991. He is a diver even now, in his seventies, and an aviator and science fiction writer whose books feature deep-diving humans and little-known underwater technologies such as supersonic cavitating torpedoes. He talks delightedly about his organization's work. "We support," he says in what seems like a well-rehearsed sound bite, "the warfighter's capabilities by pushing the boundaries of what divers can accomplish." And, more succinctly, "We enhance the envelope."

Almost a hundred years after Haldane, the EDU continues to test decompression schedules. The unit also undertakes extreme cold-water diving. The water in the EDU's wet pot—the flooded pressure chamber—can be dropped to temperatures found in places like the Arctic and Antarctic.

"It turns out that warmth is a big factor in decompression," he says. Divers who are cool but not cold at depth and who are warm during ascents and decompression tend to suffer fewer episodes of decompression sickness. The findings were so clear that at least one experiment was discontinued after a handful of the planned dives offered an indisputable conclusion in the form of far-too-frequent decompression sickness.

Echoing George Bond, John talks of the importance of controlled experiments. "Free divers and some commercial divers are guinea pigs in uncontrolled experiments," he says. They push limits, but they do not follow research protocols. They do not publish results. Since they are experimenting on themselves, they do not have to meet institutional ethical guidelines. Therefore, they can do things that the Navy would not allow, and they can and do die trying.

Regarding EDU divers, he has this to say: "They'll get hot, they'll get cold, they'll get bent, but they keep on volunteering." His tone carries more than a shade of admiration.

Regarding the ethics of human experimentation, he says that the EDU has a license to hurt people, but he is being facetious. The EDU does not have a license to hurt people. Research authorization for experiments on human subjects comes through the Department of the Navy Human Research Protection Program, known to some as the DON HRPP. The authorization requires informed consent. But requirements go far beyond that. Experimental designs have to minimize risks, and protocols include something called "termination criteria." The work can be called off—terminated—if risks outweigh benefits. And as already mentioned, the divers themselves, the experimental units, can "self-terminate," meaning not that they self-destruct, but rather that they can pull out of an experiment at any time.

But what if something goes wrong anyway? What if, for example, a diver develops symptoms of decompression sickness? This is neither the Brooklyn Bridge nor John Scott Haldane's laboratory. If symptoms appear, divers are immediately treated. All experiments are conducted in settings where prompt recompression is not only possible but routine. The treatment capabilities are, arguably, the best in the world.

This leads me to tell John about my own experience, from many years earlier, when I was nineteen years old and working in a Gulf of Mexico oil field. I was part of a crew working in about 240 feet of water when one of the divers lost consciousness. Three of us on deck hauled him to the surface. His diving helmet was half-full of vomit and seawater. He was nonresponsive and appeared, to our untrained eyes, to be dead. Working quickly, we removed his helmet and carried him—well, dragged him—to a decompression chamber. We loaded him into the chamber, and I jumped in with him.

I assumed that I would be inside with him for twenty-four hours, the time required, by protocol, to decompress a corpse. But to my surprise, at a pressure equivalent to a depth of about 90 feet, he started to come around. At a pressure equivalent to 165 feet, he was conscious. He was in pain and partly paralyzed, but he could talk. And, oddly, bubbles had formed or lodged somewhere in such a way that he was not in control of his emotions. One moment he would be angry, then laughing, then sobbing.

After some time at 165 feet, his symptoms improved. A physician was helicoptered out to the ship, and we began breathing a decompression mix with elevated oxygen levels through masks strapped over our mouths and noses.

For three days we slowly ascended together in a chamber with a diameter of fifty-four inches and a length of about seven feet.

There was no standing room and only enough space for one person to lie down at a time. That person was, of course, the victim. I squatted or curled next to him.

By day three, we were shallow enough to continue our very slow decompression on pure oxygen. At that point, we were also feeling the early symptoms of chronic oxygen toxicity. As always, it came on slowly, first as a dull ache in the chest, and then growing to a burning sensation. The victim refused to take any more oxygen, but I followed the physician's advice and continued to breathe oxygen.

While all of this was happening, our ship was steaming into the port of Venice, Louisiana. The chamber, with us in it, was lifted from the ship and loaded onto a flatbed trailer. Chains were run to secure the chamber and its occupants to the truck. We were driven, under pressure, to a hospital in New Orleans.

By the time we surfaced, I was exhausted, not only from lack of sleep and stress but also from the oxygen.

The injured diver was conscious when the chamber hatch opened. He was loaded onto a stretcher and wheeled away, soon to go back under pressure in a larger chamber staffed with trained hyperbaric nurses.

No one seemed quite sure about what to do with me. I told them about the pain in my chest, about my difficulty breathing. They took blood and urine samples and sent me home. My apartment was one floor up. With the help of my girlfriend, it took me ten minutes to climb those stairs.

"Chronic oxygen toxicity," John tells me, "is systemic. It's like other poisons. It affects the whole body, not just the lungs."

This is more than I have ever heard about what I experienced more than three decades before. And it makes sense. It is true to my memories of just how terrible—just how poisoned—I felt those many years ago.

John asks how long it took me to recover. I tell him that I was back at work within two days. He says it probably took longer than that for me to truly heal. I was paid, however, by the hour, and no one was worried about lingering chest pains or overwhelming fatigue in a teenager working in the Gulf of Mexico oil fields circa 1980. At nineteen, I was not interested in complaining. By my own assessment, I was at least as indestructible as other kids my age.

I share other sea stories with John, mostly of the bad-day-in-the-office variety. I tell him of being hauled up over the back of a diving support vessel only partially conscious after a dive to 270 feet, and of the miraculous sense of relief that overwhelmed me when I went into a chamber. I describe convulsions I experienced at around three hundred feet, three that came on and disappeared in seconds, after which I continued working, too stunned to say anything to my topside support. They were caused, I suspect, by some sort of contamination in the gas supply. I tell him of overbreathing a helmet in shallow water—of demanding more air than the helmet could deliver—to the point at which I felt starved for air and unable to catch my breath, of the temptation to dash for the surface and rip off my helmet. I repeat the story about my dreams of escaping from saturation chambers to sit in the sun on deck. I talk about pain in my shoulders that lasted for more than twenty years before slowly disappearing. All of this, with the exception of the disappearing shoulder pain, was from before I became a biologist, before I became a writer.

"You have had," he says, with no appearance of irony, "a storied career."

John, like others I talked to at the EDU, tells me about the Navy's plans to reactivate a saturation system and use it to study Gulf of Mexico ecosystems down to one thousand feet, possibly funded in part by money coming out of oil spill fines. He thinks or maybe hopes that societal pressures may eventually lead to colonies on the continental shelf and the long-term realization of George Bond's dream. Colonization would be possible at depths of six hundred feet or so. But he talks, too, of depth limits.

"Extreme deep dives," he tells me, "mess with the cell membranes and fix the limits of depth for divers."

Those limits are not a precise depth. They vary from person to person and depend, in no small measure, on a diver's willingness to tolerate uncomfortable conditions, conditions such as dyspnea—an inability to get a full breath—along with what amounts to symptoms of insanity. He tells me of talking to a French diver who had been involved in a chamber dive beyond two thousand feet, a dive on a mixture of a very small percentage of oxygen and larger percentages of helium and hydrogen. Yes, hydrogen. French divers are not known for timorousness. Yet the diver told him that he would never do it again.

Just down the road from the EDU is the Man in the Sea Museum, which is run as a nonprofit organization showcasing diving in general and the Sealab program in particular. Thanks to the efforts of those involved with Sealab itself, the museum's doors have been open for more than three decades.

The partly restored Sealab I sits outside, next to the museum's parking lot. Nearby is the Navy's first modern saturation diving system, the sort that housed saturated divers in a pressurized chamber on the deck of a ship or barge but sent them back and forth to the work site in a bell. There is also a mine-clearing device and a small submarine.

Inside, the museum houses vintage diving equipment, framed newspaper articles about Sealab, and scale models.

The museum, for someone like me, is a mecca, a magical place full of the hardware that made impossible dives possible and teeming with the stories that the hardware tells. It is the kind of place that every diver should visit.

But the museum is also tragic. Rust grows on the historic diving systems outside. The displays inside, while informative, could use attention. The overall condition of the museum seems to epitomize the way our nation feels about Sealab: indifferent.

I leave a donation and pay for a commemorative brick that will be laid in the sidewalk leading to Sealab I. But it seems to me that someone—someone with far more money than I have—needs to pry open a bigger wallet. The government and the Navy, some might think, should take more pride in their accomplishments, even if Bond's dream of colonizing the continental shelves has never quite borne fruit. At least not yet.

⌐

"Free divers and some commercial divers," John Clarke had told me, "are guinea pigs in uncontrolled experiments." The same might be said of some technical divers.

Technical divers—what many call "tech divers"—can be thought of as recreational divers who use more gear and have more training than typical scuba divers. Rather than air, they often breathe various gas mixes, including those with helium. They dive deeper, stay at depth longer, and make decompression stops, often for hours at a time. Some use ordinary scuba tanks, possibly four or five or six of them strapped to their backs and slung beneath them, and more tied to ascent lines for decompression. Others use rebreathers.

Technical divers also, as a rule, go into the water with a different attitude than typical recreational divers. They understand that things can go very wrong, and they systematically take that reality into account, limiting the possibility of disaster. A favorite word among technical divers is "redundancy." They do not count on any one piece of gear to function flawlessly. Another favorite is "planning." Technical divers often spend more time preparing for dives than they do on the bottom.

Many technical divers see themselves as explorers, diving deep wrecks and reefs and caves. They rely on experience, both their own and that of their collective community, and do not see themselves as guinea pigs.

Others, though, perhaps take pride in their willingness to experiment on themselves. There is, for example, Ahmed Gabr. In 2014, near Dahab, Egypt, Gabr descended with multiple scuba tanks strapped to his body. In about ten minutes, he reached his deepest depth, 1,090 feet, and turned around. His decompression obligation: almost fourteen hours. After surfacing, he was widely reported as saying he felt "fantastic."

There can be little argument that what he did, while incredible, was by its nature experimental. He did not respond to my requests for an interview, including emailed questions, politely worded, about whether he considered himself to be a human guinea pig. He survived unharmed, which, no doubt, can be considered a positive outcome, but he did not follow research protocols. And it is not as if there are a dozen other technical divers waiting in line to replicate his experiment.

While some in the technical diving community admire his achievement, others condemn it as a pointless non-exploratory dive that was nothing more than a long vertical swim followed by fourteen hours of decompression. Potentially deadly, the stunt, done with inadequate contingencies, with insufficient backup, relied, some say, too much on pure luck.

But if any of that is true, the same could be said of Max Nohl in 1937. And while I would not want my son to try a dive like that of

Ahmed Gabr, I would not dismiss it as a mere stunt. Gabr was, in fact, an explorer. Like Nohl, he was exploring the frontiers of his own abilities. He undertook an uncontrolled experiment to determine his ability to reach 1,090 feet more or less on his own, without the support of a saturation system or an on-site pressure chamber, and he succeeded. But few—possibly including Gabr, since he has not been back to 1,090 feet since his 2014 dive—would argue that there is much of a future in his approach.

Through a network of old friends and colleagues, I contact Steve Porter. Like me, he was diving in oil fields by the time he was nineteen years old. Unlike me, he had both the opportunity and willingness to participate in an extremely deep experimental dive in a chamber at Duke University in 1981. With two other volunteers, he was pressurized to what amounted to 2,250 feet below sea level. They lived in a steel can for forty-three days, residing briefly at pressures close to those found a half mile underwater.

"We started," Steve tells me, "in a chamber called the Golf Ball, a seven-foot-diameter ball on top of a six-foot-diameter vertical cylindrical portion, separated with an aluminum deck between the sphere and cylinder. We lived and took most tests in the top. The deck opened up to provide a ladder to the bottom section, where we had a crapper and the cardiopulmonary test setup. The cardiopulmonary tests were incredibly invasive, with venous and arterial blood sampling and hundreds of inhaled and exhaled gas tests."

The men pedaled an exercise bicycle, used tweezers to move tiny ball bearings into a narrow tube, dropped washers onto pegs, and played with wrenches and bolts. All of their urine went to laboratories.

They tolerated one another in a space no larger than a typical American walk-in closet.

"Our day started every morning with a venous stick and four samples of blood before breakfast," he recalls. "We centrifuged the samples to get the cells out of the serum and sent the serum out to the hematologists."

And there was the challenge of breathing itself. At that depth, breathing requires a certain level of concentration. "We became," Steve tells me, "fully obligatory mouth breathers." Their atmosphere was too dense, too thick, to move conveniently through nasal passages. Eating required chewing and swallowing between breaths.

The most tender of steaks, with all of the air squeezed out of them, became tough. The creamiest bowl of ice cream was, at 2,250 feet, squeezed into a thimbleful of flavored ice.

The pressure affected their minds. "Dreams...Yikes, the dreams were amazing. Technicolor, very vivid, very realistic, running the gamut from horror to incredibly erotic. Never had any like them before or since."

And there was straight-up madness. "We had one subject go stark raving crazy," Steve tells me of one of his extreme dives. "We had to hit him with enough lithium to make an elephant sleep."

Through the chamber's intercom, a psychiatrist outside gave Steve the technical name for the symptoms he was seeing. "The shrink said, 'Yeah, I think that's what it is, and I don't know how you guys are doing it,'" Steve recalls. "'I can't even stay in the same room with nuts like this.'"

The decompression schedule had to be made up as the divers ascended. "It turns out the mathematical models used to develop normal decompression tables were completely useless," he says. Later in life, he says, he learned about fluid dynamics. Normal decompression models assume that fluids—like blood, for example—are incompressible, and at normal diving depths, for all practical purposes they are. But at 2,250 feet, fluid compressibility might matter.

Their first incidence of decompression sickness—their first hit, as Steve calls it—occurred before their ascent reached two thousand feet. And the hits kept coming. Some they described as "niggles," a kind of minor irritation that felt as if it was moving through the body before disappearing, as though a bubble was making its presence known as it traveled along a blood vessel. Other symptoms were more severe.

Each hit delayed the decompression twenty-four hours. The ascent was halted while the symptoms cleared.

At one thousand feet they moved into a larger chamber, one that had been unavailable to them at their deepest depth because it was not rated to withstand the pressure. The three men moved from an area about the size of a walk-in closet into an area about as big as a child's bedroom.

"We had hits all the way up," Steve says.

As they got shallower, the temptation to ignore the hits grew. "When you're almost home and in your lady's arms," he says, "it takes monolithic discipline to call a hit and hold the team in the chamber for another day."

About a decade later, in 1992, divers went deeper. These were the French divers that John Clarke mentioned during our conversation at the Experimental Diving Unit, the team breathing a mix with a small amount of oxygen and larger amounts of helium and hydrogen. Just as Steve and his colleagues had been, just as many of Haldane's and Bond's experimental divers had been, the French divers in 1992 were in a chamber. They reached a pressure equivalent of 2,300 feet, slightly deeper than Steve's depth. Since then, over two and half decades have passed, but no one has bested that record.

So I ask Steve if that was it. Have humans reached the practical limit for diving?

"I believe we've reached an economic but maybe not a physiological limit," he says. With the right economic pressures, further research might illuminate the ways in which pressure changes the way people think. Medication might block some of the symptoms. Better screening of divers could weed out those who might not be able to cope with the conditions of extreme depths. Steve thinks that at least some people might be able to function at, say, three thousand feet.

Or maybe not. Experimentation with deep diving, with exposing human beings to extreme depths and pressures, has essentially ended. Surface-based saturation systems support working divers in oil fields and elsewhere, but dives beyond a thousand feet remain rare. And little has come of George Bond's dream of underwater colonies. But for a few scattered underwater habitats in relatively shallow water, including some tourist operations, the dream of deep sea colonization by conventional divers, by humans exposed to the pressure of the sea, has been, at least for now, set aside. The ability to go deeper while performing useful work, the ability to actually live on the continental shelf, might require pressure-proof hulls and robots.

Chapter 5

SUBMERGED

Phil Nuytten is in his seventies and has been around diving in one capacity or another since high school. He has cleaned boat bottoms, salvaged ships, and worked in oil fields. He was among the first wave of civilian saturation divers.

We meet at his business in Vancouver, British Columbia. "Pressure," he tells me, "is the enemy."

Not only has pressure killed and injured friends of his, it has damaged his own hearing—not because of noise from the movement of gas or the incessant din of compressors familiar to so many divers, but because of an errant bubble not appropriately off-gassed. Put another way, he may have suffered from a vestibular hit, a surprisingly common form of decompression sickness in those willing to breathe exotic mixtures of gas and to switch from one to another in an effort to shorten decompression times. And so, yes, pressure is the enemy.

Phil believes that divers, especially those going deep and staying for long periods, might be better off if they were armored in such a way as to make the breathing of compressed gas unnecessary. Humans encased in rigid cocoons could be lowered to great depths without directly exposing their bodies to the dangers of pressure.

He is far from the first to consider such an approach. One of his predecessors was John Lethbridge, with his Lethbridge diving machine, first used around 1715. "It is made of wainscot," wrote Lethbridge, "perfectly round, about six feet in length, about two foot and a half diameter at the head and about eighteen inches diameter at the foot, and contains about thirty gallons; it is hooped with iron hoops without and within to guard against pressure."

"Wainscot" is, of course, wood. Lethbridge's face rested in front of a glass panel allowing him to see out of his wainscot barrel. His arms poked out through holes in the barrel, with leather seals around his biceps keeping the water out. This diving "suit" was weighted so that it would sink and was suspended from lines so that he or anyone else using the contraption would hang horizontally in the water. Lethbridge claims to have made dives as long as six hours, broken up at intervals so that he could be hauled up to have fresh air flushed through the wainscot shell.

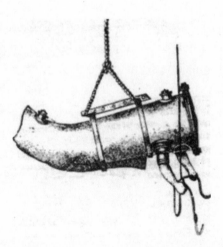

A version of the early atmospheric diving suit built and used by John Lethbridge. No drawing or photograph of Lethbridge's suit exists, but the suit illustrated here, probably used in the nineteenth century, was based on the detailed description provided by Lethbridge in the September 1749 issue of Gentleman's Magazine.

The leather arm seals would have limited his working depth, but other inventors followed with entirely pressure-proof diving suits, essentially personal submarines that looked something like medieval armor or humanoid robots, with bendable joints at the elbows, shoulders, knees, and hips. There was, for example, the Carmagnolle brothers' suit of 1882, with twenty-two rotating joints and twenty-five two-inch glass ports in the helmet, all of which contributed to an unfortunate tendency to leak. There was the U.S. Navy's own lobster-clawed diving armor of 1913, now on display at the Man in the Sea Museum in Panama City Beach, Florida. There was the eight-hundred-pound Tritonia of 1930, which dived on the sunken RMS *Lusitania* at a depth of about three hundred feet before being retired in 1937. And a host of others.

Two divers prepare to explore the wreck of the RMS Lusitania *in 1935. One (at left) wears the Tritonia atmospheric diving suit, and the other wears the standard diving dress, almost unchanged since its introduction by Siebe.* (National Oceanic and Atmospheric Administration, Department of Commerce, and OAR/Nation Undersea Research Program [NURP])

These suits came to be known as atmospheric diving suits because the diver inside was exposed to no more than one atmosphere, the pressure found at sea level, on the deck of a ship, or on the dockside. The diver breathed as though he or she were on the surface. There was no exposure to pressure and therefore no decompression requirement, no narcosis, no oxygen toxicity, no requirement for heliox. Divers could descend and ascend without decompression. For that matter, they could descend and ascend without getting their hair wet.

In a TEDx Talk, Phil Nuytten described himself as a man who "builds contraptions out of parts that do not yet exist." Surrounded by the fruits of his labor in his Vancouver facility, he repeats that self-description to me but also says that he thinks of himself as a blacksmith.

He tells me about a group of Chinese scientists and engineers who were interested in his suits, and specifically in the joints. To these Chinese, joints that do not lock up at depths of a thousand feet or so, where the pressure pushes the components together, seemed impossible. After some consideration, they insisted that his approach would not work. They implied, he told me, that they thought he was lying. He told them that they were wrong and the joints were already functional.

"Pressure adds risk," Phil says, "but it also adds cost." A saturation dive to one thousand feet, he reminds me, requires about one day for pressurization and ten days for decompression, which does not allow much time for actually getting any work done. The helium for a one-thousand-foot saturation dive might cost more than $100,000. The support vessel might cost $75,000 a day, whether the divers are working or compressing or decompressing.

An atmospheric diving suit—a suit like his own Newtsuit, or his more modern Exosuit, both of which have been described as "submarines that you wear"—might be able to do the same job for $50,000 and one day of ship time.

He was once quoted as saying that he would eventually get the price of his atmospheric diving suits down to about $25,000. For that price, I tell him, I would like to buy two.

"What I actually said," he tells me, "is that I would like to get the cost down to something like that of a sports car." The current cost of one of Phil's suits is $750,000, roughly equivalent to that of a very high-end sports car.

In the world of deep sea exploration and innovation, Phil Nuytten is a widely admired individual, but few share his zeal for atmospheric diving suits. Even though the person inside is not exposed to pressure, the suit is. In principle, the Chinese scientists and engineers were correct: at some point extreme pressure prevents the joints from bending.

And the suits tend to be what most people would think of as heavy and a little bit clunky. The diver inside cannot rely on his or her sense of touch.

When talking to various experts, I often hear the suits described as "another tool in the toolbox," as something that might be useful in very specific circumstances, for certain jobs. One leader in the field, whom I will not name here as a matter of discretion, calls atmospheric diving suits "very expensive toys." Some say that divers in atmospheric suits cannot do anything that a robot cannot do at a far lower cost.

But others—and I include myself in this group—say, "I wish I could afford one."

⌐

While coming to Vancouver for no other reason than to look at Phil's suits would have been well worth my time, I was drawn there to hear, straight from the inventor himself, about what he calls Vent Base Alpha, his vision for an underwater colony.

At Vent Base Alpha, colonists would live far below the surface near the geological features that have come to be known as smokers, a kind of hydrothermal vent, something like a hot spring on steroids that occurs only at great depths. Smokers spew groundwater at temperatures approaching one thousand degrees Fahrenheit, water coming up from deep in the earth, from places where it is warmed by the planet's inner furnace. And around them unique communities of deep sea clams and squat lobsters and giant crabs thrive.

On the surface, water as hot as that coming from smokers would be steam, but under the pressures found far down on the continental shelf, it remains a liquid. That liquid carries dissolved copper, zinc, cobalt, silver, and gold. Around the vent, as temperatures drop, valuable metals precipitate, forming rich ores. Vent Base Alpha's colonists, Phil believes, could turn a profit.

They would not mine the minerals in any conventional sense, which would require that they disturb the ecosystem, dig up the ore, and trash the homes of deep sea clams and squat lobsters and giant crabs. Phil may be in his seventies and a man who has made his living through innovations that have helped, among others, the military and the oil and gas industry, but he loves the sea and believes in sustainability. Strip-mining is simply not his thing. His colonists would not disturb the seabed to get at its minerals, but instead they would take them directly from the hot water.

Vent Base Alpha's energy would come from the same source. As Phil sees it, the colonists would use a Stirling engine, a device relying on the temperature differences between the vents and the much colder surrounding seawater, to generate electricity. That power would then be used to run, among other things, lights, which could in turn be used to grow crops. Colonists would eat their own vegetables, along with sustainably harvested seafood.

And all of this would be done using small submarines and atmospheric diving suits. No one would breathe helium or hydrogen. No one would lose their hearing to a tiny bubble, to a case of decompression sickness.

⌒

Investors are not, at this time, lining up to fund Vent Base Alpha. Phil's dream of deep sea colonization may be before its time. But just south of his facility, across the border in Washington State, lies Naval Base Kitsap, home to a fleet of U.S. Navy ballistic missile submarines. The Navy personnel roster includes something like 26,000 active duty submariners. And in a book called *Power Shift*, recommended to me by Don Walsh himself, a retired submariner estimated that American ballistic missile submarines have spent, cumulatively, just shy of 605 years underwater. It is reasonable to assume that these submarines each carried, on average, at least one hundred men. In total, American ballistic missile submariners have accumulated more than 60,000 person-years underwater. This estimate does not include sailors staffing fast attack submarines, or those on nuclear submarines operated by other nations, or the men and women who live and work in the diesel-electric submarines owned by nations such as Holland, Japan, and Iran. And it does not include time spent on submarines after the estimate was made, which was more than ten years ago.

Phil's dream to see the deep sea colonized using one-atmosphere technology is, it seems, already a reality. Just not in the way he envisions.

⌒

I talk to Geoffrey Moss, a civilian employee who holds the title "Science and Technology Advisor at Commander Submarine Forces."

"Outside of the international space station," he tells me,

"submarines are the most complex things human beings have ever built." This surprises me. A vessel capable of going underwater under extreme pressures, staying there and working autonomously for months at a time, maintaining aboard an extraordinarily destructive collection of ballistic missiles that can be fired from beneath the surface, supporting roughly 150 crew members, and doing it all very quietly to avoid detection by potential adversaries, seems to me to be far more complicated than a machine that drifts through space in a more or less fixed orbit.

Submarines were not always so complicated. Take, for example, that of Cornelis Drebbel, born in Alkmaar, Netherlands, in 1572. After moving to London in 1604, he worked for King James I developing various inventions. One was a supposed perpetual motion machine. Another cooled part of Westminster Abbey. And a third was his submarine, the world's first, or purportedly so, built sometime before 1623. The hull was made from wood and leather. Oars extended outside, presumably sealed by leather and pitch and rowed by men who sat inside, working under Drebbel's direction.

Drebbel offered demonstrations on the Thames, running submerged (or so people said) from Westminster to Greenwich. In 1660, Robert Boyle himself wrote of Drebbel's submarine, calling it "a vessel to go under water, of which tryal was made in the Thames with admired success, the vessel carrying twelve rowers besides passengers." Although some accounts suggest that the submarine pulled air in from the surface through what might be best described as snorkels, others, including Boyle, suggest that it was entirely sealed.

Boyle wondered why the rowers and the passengers and the inventor himself did not die from asphyxiation, and he realized that Drebbel had in his possession some sort of chemical mixture that restored the air. "When from time to time," Boyle wrote of Drebbel, "he perceived that the finer purer part of the Air was consumed or over-clogged by the respirations and steams of those that went in his ship, he would by unstopping a vessel full of this liquor, speedily restore to the troubled Air such a proportion of vital parts as would make it again for a good while fit for respiration." Although exactly how it was done remains unknown, it may have relied on an early version of a carbon dioxide scrubber, the same sort used by Max Nohl in his dive to 420 feet in 1937, by Navy divers on Operation Ivy Bells, and by some technical divers today. And also, for that matter, in certain situations as a backup system in modern nuclear-powered submarines.

Since Drebbel's time, thousands of submarines have come and gone, some more famous than others, some entirely unsung, and many dangerous almost beyond belief. Among the famous was the American *Turtle,* a wooden egg-shaped thing with room for a crew of one built to attack British ships in 1775 but never successfully deployed. Another was Robert Fulton's submarine. Fulton would later grow rich commercializing steamboats, but in 1800 he built a twenty-one-foot-long submarine with a copper-sheathed hull. It was propelled by hand cranks underwater and by a sail on the surface.

Among the most dangerous submarines was Wilhelm Bauer's *Brandtaucher,* or "burn diver," built in 1850 with a riveted steel hull, a treadmill linked to a propeller, and sliding weights used to trim the vessel's tilt during dives. On the *Brandtaucher's* second dive, the weights slid forward unexpectedly and sent the craft steeply downward, nose-first into the mud, leaving it and its crew stuck on the seabed in sixty feet of water. Because of the pressure, the hatch was pinned closed. Everyone was trapped inside. Bauer told his men that they must flood the vessel, allowing seawater in until the pressure inside equaled that outside, at which time they could open the hatches and swim to the surface. He may or may not have told them to exhale forcefully on the way up, that the pressurized air in their submarine and in their lungs would expand during their ascent and would kill them if they did not exhale. In any case, they survived.

The Civil War saw the deployment of the Confederate submarine *H. L. Hunley,* propelled by hand cranks linked to its propeller shaft. The *Hunley* had an unfortunate habit of sinking and killing her own crews even during training, but she eventually sank the Union's USS *Housatonic,* making her the first submarine to destroy a wartime adversary. But in doing so, she sank herself and killed all eight of the sailors aboard, after which she stayed on the seabed until she was raised almost a century and a half later in a salvage effort funded by none other than novelist Clive Cussler, known for his stories of underwater heroism.

In Cartagena, Spain, I visit the Museo Naval, chiefly to see Isaac Peral's submarine, named, appropriately, the *Peral.* She is often described as the world's first diesel-electric submarine, meaning that she had a diesel engine for propulsion on the surface and to charge batteries, and a battery-powered electric engine for propulsion while submerged.

Most of the museum's signage is in Spanish and English, leaving no doubt about Spanish national pride. Example: "Despite other former attempts, with varying results Spain places itself as one of the pioneers in the creation of the submarine weapon. Lieutenant Isaac Peral, distinguished naval officer born in Cartagena, is considered to be the father of the modern submarine. He designed and even built a model of a submersible vessel which, in 1880, was already equipped with the main instruments which characterize this type of vessel in the 20th Century."

Another panel claims that the *Peral* was the first submarine to launch a torpedo underwater.

Spanish pride extends from *Peral* the submarine to Peral the man. "Local authorities," a panel explains, endearingly using the present tense for a man long dead, "contest for Peral's presence in order to open streets and monuments. Several guilds pay tribute to him: grocers invite him to try their wines, so they can ask him to sign casks, bullfighting business men host bullfights in his honour, painters portray him, sculptors carve him busts, poets write him poems."

His submarine, or what is left of her, stands in her own hall. She is seventy-two feet long and just over nine feet in diameter, a long tapered tube of a thing, with lines of rivets holding her steel plates in place. She looks like she might have been featured in a 1950s science fiction movie.

A line of gawkers, mostly Spaniards, wait to climb a short flight of stairs, at the top of which they can peer inside the submarine. I wait my turn, and five minutes later, looking into Peral's machine, I am not sure if I am impressed or indifferent. What I see, both inside and out, looks like a weighted boiler with propellers. I am left momentarily underwhelmed.

But then I remember that this machine's thirty tons of batteries allowed her electric motors to push her along at ten knots beneath the waves and that she could stay at depths of at least thirty feet for more than an hour at a time, all in 1888, only eight years after Edison patented his electric lightbulb, just seven years after Sitting Bull surrendered in Montana, and eleven years before Henry Ford launched the company that would, after a few false starts, allow Americans to retire their horses. Staring into the submarine at the top of the stairs, I overstay my welcome and am nudged along by a group of teenagers who are laughing and talking too quickly for me to comprehend what they are saying.

In World War II, German diesel-electric submarines wreaked havoc with Allied shipping in the Atlantic, while American diesel-electric submarines did the same with Japanese shipping in the Pacific. But there were also British, Dutch, French, Chilean, Brazilian, Canadian, Soviet, and Italian submarines. There was, for example, the notoriously successful Italian *Leonardo da Vinci,* credited with the sinking of seventeen ships before she herself was sunk, with all hands, on May 23, 1943.

World War II submarines spent most of their time on the surface, running their diesel engines for propulsion and to charge their batteries. Submerged, they used electric motors to turn their propellers. For the most part, they submerged only to attack or to evade. At least some were capable of operating deeper than seven hundred feet.

They were dirty, uncomfortable, and crowded not only with men but, when leaving port, with supplies and explosives. "Hot sheeting"—the sharing of a bunk by crew members on different work shifts—was a common practice, as was smoking. Fresh water for bathing was more than scarce.

Despite the close space and enclosed atmosphere, these vessels were killing machines. For example, Germany's *U-156* sank the British troop carrier *Laconia* on September 12, 1942, killing 1,621; the USS *Sturgeon* sank the Japanese troop carrier *Toyama Maru* on June 29, 1944, killing 5,400; and the Soviet submarine *S-13* sank the German evacuation ship *Wilhelm Gustloff* on January 30, 1945, killing 9,343.

All told, German submarines sank more than 2,700 Allied ships during the war. American submarines in the Pacific were responsible for sinking more than half of Japan's warships and three-quarters of her merchant ships, making submarines in the Pacific a critical deciding factor in the war.

The submarines were deadly not only to their enemies but also to their crews. "Of all the branches of men in the forces," Winston

Churchill once said, "there is none which shows more devotion and faces grimmer perils than the submariners." Of the 40,000 or so who served aboard German submarines during the war, 28,000 died. Something like one in five American submariners were lost during the war. The dead remain on what survivors euphemistically call eternal patrol.

The world's first nuclear-powered submarine, the USS *Nautilus*, went to sea on January 17, 1955, famously radioing to shore, "Underway on nuclear power."

A point of clarification is needed here. Nuclear-powered submarines, like their diesel-electric predecessors in World War II, rely on electric motors. But instead of generating electricity using diesel engines that run only on the surface, where unlimited oxygen is available, they generate electricity using steam turbines that rely on the heat produced by nuclear fission. Oxygen from the air is not needed, even for breathing. With electricity, it is a simple matter to make enough oxygen from water to support the crew. And the fuel for nuclear reactors does not disappear in a matter of hours and days, as it does in diesel submarines. A small amount of fissionable fuel will last for years.

Nuclear submarines can stay submerged until they run out of food, or, as some naval officers have told me, "until it is time to reenlist the crew."

In Washington, at Naval Base Kitsap, not far from Seattle, I meet forty-three-year-old Matthew Chapman, commander of the USS *Alabama*. I want to talk to him about the realities of what are seldom seen or described as mobile undersea colonies, despite the fact—and Jules Verne would concur with me—that this is exactly what they are.

Commander Chapman and his submarine will be leaving both their home port and the surface of the ocean within the next seventy-two hours, bound for an undersea voyage that will last two to three months. Even so, when we shake hands in his office, he strikes me as a man who is more relaxed than, say, the average yachtsman preparing to leave the dock. As mellow as, say, a surfer sitting on a beach. I ask if he really has time for an interview. Would his time be better spent supervising his boat and crew?

"My crew," he says, smiling, "has it covered."

His submarine, for obvious reasons, is not available for a civilian tour. He seems to regret this reality, and while I would love to look around the *Alabama,* or maybe dive with her and her crew for at least a few days, I am happy to have been granted an interview. The Navy is not, after all, in the business of entertaining writers.

Many of my questions regard technical matters. I ask, for example, about crush depth. The actual figure is classified, but he tells me that his submarine can operate to depths of at least four hundred feet, and he is aware of public sources that claim it can operate at eight hundred feet. But what I really want to know is what would happen if he descended beyond crush depth. Would valves start to fail? Would water start pouring in around the propeller shaft? He seems to find my question amusing, the result of my watching too many World War II submarine movies. In reality, he says matter-of-factly, the hull itself would fail. It would, apparently, flex and crack, and then, broken, it would allow the sea inside.

Some of my questions cannot be answered. He cannot tell me how thick the hull is, for example, or if some sections are thicker than others. Such information, he suggests, might give an adversary an advantage in aiming a torpedo or planting explosives.

Other queries draw similar responses. He is not being evasive. He just does not want to see either one of us arrested.

He is more forthcoming about his crew. As his submarine is prepared for sea, he has to know that everything of importance is working. "Everything" in this context most certainly includes the crew.

"Twenty years ago," he tells me, "the Navy was tougher. But now it's all about coaching. Sailors need motivation and recognition. There is more mental health help than in the past."

There is no formal psychological screening for submariners. They volunteer, and if they are not weeded out during training, which is a real possibility, they are deployed. In general terms, good submariners are resilient and adaptable team players who do not—and this point should be obvious—suffer from claustrophobia.

But there are longer-term psychological challenges at play. "New crew," he says, "might not be conscious of the psychological transition needed when we go to sea and when we return." They may not think through the realities of regularly switching from one way of life with one set of responsibilities to another way of life with another set of responsibilities, from submarine to spouses and children.

The crew, especially the new recruits, are watched and educated through what he calls "a pyramid of supervision." Commander Chapman watches his officers, who watch their chiefs, who watch their men. Informal mentorship is offered when needed. Issues are reported upward.

Those aboard understand the difference between chronic and acute stress, between the wearing down of a sailor's psyche during weeks of routine and the intensity of an emergency lasting a matter of minutes or even seconds.

While the *Alabama* is at sea there are daily discussions about morale.

Commander Chapman knows the roughly 150 men who make up his crew by name. "By last name," he says, "like in high school gym class."

I ask what kinds of things submariners take with them for sixty to ninety days underwater. He tells me they bring portable electronics, including Kindles for reading, and a few actual books, and cribbage boards. "Cribbage," he says, "is a big game aboard."

One sailor departs from the surface with a small guitar.

I ask Commander Chapman what he talks about at cocktail parties. "What do you say after you've told an acquaintance that you command a ballistic missile submarine?"

He likes, he says, to talk about the mission of strategic deterrence. He discusses the challenge of defining success as "not having to do anything." By which I take him to mean "not having to launch a missile" or "not having to participate in Armageddon."

"The twentieth century," he says, "saw twenty million war deaths. Strategic deterrence prevents more deaths. We have saved lives and protected prosperity by avoiding another world war. We no longer have to rebuild the world every few decades."

Success, then, is monotony. It is cruising around beneath the waves unheard and unseen and, in terms of launching missiles at enemies, entirely inactive.

Before I leave, I give him a book for the ship's library. It is not a book I have written, but is instead a full-color coffee-table book produced by a company that manufactures small submarines sold, for the most part, to the owners of megayachts. He takes a quick look at a few of the photographs. The submarines—more correctly called submersibles—sport leather seats behind transparent domes. The occupants can see outside.

Commander Chapman, in charge of one of the largest and most sophisticated submarines the world has ever seen, but one with neither windows nor leather seats, is openly delighted.

"I'd love to have one of these things," he tells me.

"Me, too," I confess, "but a basic model goes for well over a million dollars."

For now, Commander Chapman will have to remain content with the *Alabama*, which cost, when it was built in 1984, something like two billion dollars, even though it has neither leather seats nor a transparent dome.

⌐

I am invited—or, really, I invite myself, through the Navy's polite and efficient public affairs bureaucracy—aboard the USS *Tennessee*, another ballistic missile submarine, as far as I know similar in almost every way to Commander Chapman's *Alabama*. She is, on the day of my visit, secured to a dock in Kings Bay, Georgia, undergoing scheduled maintenance.

My escort and I spend thirty minutes negotiating security checkpoints staffed by serious-demeanored sailors armed with assault rifles. A public-address speaker repeatedly warns that unauthorized visitors are not allowed. "Deadly force," the person or recording on the speaker announces, "will be applied." I stick close to my escort.

After flashing identification badges one final time, we are confronted with a sliding metal door at the dockside itself, a door that is raised and lowered when a guard wearing body armor calls "up" and "down" into his radio. Our turn to pass through comes without fanfare. The door slides vertically open. We enter.

And there, on the other side, the *Tennessee* becomes suddenly and imposingly visible. She is no museum piece leaving me wondering whether I should be impressed. She is one of the nation's eighteen Ohio class submarines, stretching 560 feet from bow to stern. The Navy admits that she can travel submerged at speeds of twenty knots, but of course she can go faster. Internet sources give her an additional five knots, but even this could be an underestimate. Under way, she carries 15 officers, 140 enlisted men, and 24 ballistic missiles, each of which holds up to 12 nuclear warheads.

We board the submarine and immediately descend down a spiral staircase. It is a temporary structure, installed for use here at the dock. We walk through the vessel, looking at systems, talking to submariners. When it comes to interior decorating, battleship gray appears to be the Navy's favorite color, followed distantly by black.

It strikes me that I am the oldest man aboard. Unlike most of those around me, I am older than even the submarine herself, which was launched on December 13, 1986. But I can note with some confidence that I am in better shape than many of these young men, which is not to say that I am especially fit. The life of a submariner is one of long hours, whether at the dock or at sea. Working, eating, and sleeping take precedence over recreational activities, including exercise, especially when under way. And despite the size of this submarine, there is no room for anything resembling an actual gym.

I look at a pipe running overhead with the label TD, for "trim and drain." I am told that it moves ballast water around the submarine in a manner that keeps her stable, that tilts her and levels her, in principle like the sliding weights of Bauer's *Brandtaucher*, but in practice leaves her not at all apt to send the submarine headfirst into the seabed.

I look at a sign that says THINK QUIET in white block letters against a blue background, striking not so much for the message it carries as for the fact that it is neither gray nor black.

Near the command center of the submarine, someone has tied traditional decorative knot work around a post, attractive and interesting and anachronistic in this businesslike, humorless diving machine. We walk past batteries and missile tubes and a Fairbanks Morse diesel engine sporting twelve cylinders with twenty-four pistons, the submarine's auxiliary power, to be used on the surface if the reactor is out of service. We pause in a room full of computers where crewmen sit staring at screens.

"These computers," says an enlisted submariner, aged perhaps twenty-three or twenty-four, "were installed in 1986." I take this as a joke, but I am not sure. It is possible that he believes it to be true, even though the computers look, to my eye, no more than maybe fifteen years old. When I ask him what they are for, he replies, simply, "Missile control," a statement that sends chills down my spine.

I ask about noises when the submarine descends. Does it groan under the weight of the sea? It does not. There is, I am told, hardly a sense of motion. In normal operations, the submarine stays essentially horizontal in the water column.

But sometimes the more experienced hands will tie a string across a compartment, drawing it taut, and the younger guys, watching it, will see it sag slightly as the submarine sinks into the cold darkness of the sea. Or so they tell me.

American taxpayers need not worry that the government is profligately spoiling submariners. The officers' mess seems cramped, just big enough for elbow-to-elbow dining, and it also serves as a briefing room. Its table, if necessary, can be used for surgery. The captain and his second-in-command share an adjoining bathroom. Their cabins are reasonably spacious, in part because their bunks fold up against a bulkhead when not in use. The remaining officers, who have two showers among them, are in shared accommodations. More than a hundred enlisted men share six or seven showers and sleep in bunks stacked three high. They do their own laundry, with groups granted access to washing machines and dryers on a particular day each week. Regarding onboard food, an enlisted submariner describes it as "the best that anyone can expect of a cook using what the Navy issues."

This is not Captain Nemo's submarine. There are no luxuries here.

In a short conversation with a junior officer, I mention that conditions aboard seem to be on the Spartan side, perhaps to a fault. He becomes defensive, spitting out words such as "adequate" and "comfortable," from which I conclude that he may have grown up with six brothers sharing a single room with bunk beds lining the walls. But I admire his defensiveness. I myself, as a diver working in the oil fields in the 1970s and 1980s, sometimes lived under conditions far worse, especially during saturation dives. Yet at the time I considered them adequate and comfortable, and I, too, became defensive when anyone suggested otherwise.

Everyone I talk to has the air of those accustomed and thoroughly resigned to routine. These are by-the-book guys, doing their duty. None of them seem particularly interested in what they might see out the window, if there were a window, which, of course, there is not. But I do inquire about the things they bring with them, just as I asked Commander Chapman what his crew brought along on patrols.

"Not much," one says. "We don't have the space for much."

"My PlayStation," I hear on three separate occasions.

"Books," one man tells me. "I like to read nonfiction." But then he names an author I do not recognize, and later I discover that what he calls nonfiction includes characters capable of various kinds of magic and aliens with unusual shapes and powers.

The submariners seem surprised by my interest in their breathing gases. They use electrolysis to make all the oxygen they need directly from seawater. In an emergency, they can make more by burning chlorate candles. They remove carbon dioxide with what amounts to a kind of molecular sieve, but as backup they have chemical scrubbers, too, the same sort used aboard Trieste in 1960, by divers wearing rebreathers, and possibly by Drebbel.

The young man explaining how carbon dioxide is removed from

the atmosphere interrupts our conversation when a chief petty officer appears in the adjacent corridor. "Looking good there, Chief," the man calls out, smiling broadly. "Looking younger."

"It's the fresh air and vitamin D," the chief responds, his tone upbeat and familiar, a tone I would associate with old friends encountering each other after a prolonged absence. But the chief is busy. He does not even break stride.

I ask several people if they know any good submarine jokes. They do not. Or, at least, they do not know any good submarine jokes that they are willing to share with an older stranger, a civilian with a pad and pencil.

Here is a surprise: the crew, at sea, has occasional access to short email messages, limited to a few words at a time, both sending and receiving. What kinds of messages are exchanged between submariners, gone for months at a time, and their loved ones ashore? "About that argument we had just before I left? I was right, so screw you." Or, "All is forgiven. I hope likewise." Or, more likely, "All good here. Staying sane. Miss you."

It strikes me that one of the men I have been talking to, an older sailor by the standards of this submarine, placing him in his late twenties, seems depressed and neither especially interested in submarines nor proud of what he is doing. I ask if he has heard of Drebbel and Peral and Don Walsh. He claims to have heard of the first two, but not of Walsh. When I tell him about *Trieste,* he becomes engaged, and when I tell him that Walsh, at the time of his most famous dive, was twenty-seven years old, he becomes wide-eyed. But when I describe the bathyscaphe as a gasoline-filled submersible blimp, he seems skeptical.

I change the subject, asking once more for a submarine joke, but he has nothing to offer, and my tour is over. With my escort close at hand, I pass through the vertical steel door, deeply impressed by both the machine and the young people running her. My mind is awash with new impressions about what it might be like to live in a nuclear submarine, knowing that I could never fit in, that I would be incapable of adjusting to the routine, but at the same time wishing above all else that I could have convinced the Navy to give me a little ride, just a day or two or three in one of their machines, a short visit in one of their mobile undersea colonies.

As I write this chapter in November 2017, tragedy unfolds in the waters off Patagonia. Authorities tell the public that the Argentine diesel-electric submarine *San Juan* disappeared two days earlier. A search is under way. There is speculation about how long survivors might last trapped below. It depends, in part, on how fresh the air was when she last submerged. There is speculation about how deep the water might be where she sank, and reasonable hope that it was shallow enough for the submarine to have settled onto the seabed well above its crush depth. At first there is at least the possibility that the submarine will be found and that at least some of the forty-four men onboard will be rescued.

As I hear the news, I envision men burning the kind of oxygen candles I had seen aboard the *Tennessee,* breathing through masks with personal carbon dioxide scrubbers, lying in their bunks or sitting, waiting in the dim glow of emergency lighting, doing their best to control their breathing, to remain calm. But after fifteen days, the rescue mission becomes a recovery mission. If the submarine sank in water deeper than its crush depth, the men would have died when the hull failed and the cold Atlantic found its way inside. If the submarine sat in water shallower than crush depth, the men would have perished slowly, perhaps holding on to hope as, one by one, they fell into unconsciousness.

Other submarines, in peacetime and in wartime, have lain

crippled on the seabed with survivors aboard. There was, for example, the Russian nuclear submarine *Kursk*, 505 feet long, carrying 118 men. With her double hulls and nine watertight compartments, she had at least the potential to survive a direct hit from an enemy's torpedo. But on August 12, 2000, during a training exercise, the fuel of one of her own torpedoes leaked and exploded. That torpedo was more than thirty feet long and weighed, before it exploded, five tons. Two minutes and fourteen seconds later, as recorded on various underwater listening devices, there was an almost simultaneous series of as many as seven additional explosions, each representing another torpedo exploding. Or so said later official reports.

The blasts and their immediate aftermath killed 95 of the 118 men aboard. That left 23 survivors. They gathered in the relatively intact Compartment Nine. The explosions had left the submarine entirely crippled, lying on a clay seabed in 354 feet of water. The men were trapped in water shallower than their submarine was long, but there was no reasonable way to escape.

Oxygen levels dropped, carbon dioxide levels rose, temperatures cooled, and water contaminated with various chemicals slowly rose in Compartment Nine. The submarine's batteries were drained of power, leaving the survivors in darkness.

But the men had reason for hope, at least at first. They had sunk in the midst of a massive training operation, surrounded by other ships and submarines. That hope, though, would not last.

When oxygen levels drop, the brain slows. When carbon dioxide levels rise, breathing becomes strained and the head aches. Cold numbs not only the skin but also the mind. And there was the darkness. Nevertheless, the men did what they could to improve their chances. There was, remarkably, no panic aboard the *Kursk*.

"It's too dark here to write," recorded one of the survivors on the page of a logbook, "but I'll try by feel. It seems like there are no chances, 10–20%. Let's hope that at least someone will read this." He listed the names of the other survivors. "Regards to everyone," he wrote in this most desperate of situations, and, inexplicably, "No need to be desperate."

On the other side of the same page were a few words to the writer's wife: "Olichka, I love you. Don't suffer too much."

Above the *Kursk,* after initial delays and confusion, a rescue attempt was under way. But it was too late. The men, in the end, did not die from lack of oxygen or carbon dioxide poisoning or hypothermia, but from another explosion. The contaminants in the water, at least some of which were volatile, had permeated the remaining air in the submarine, and something—possibly a chemical cartridge intended to renew the atmosphere—offered a source of ignition. In the end, no one aboard the *Kursk* survived.

Cartographers once labeled vast parts of the globe *nondum cognita,* or "not yet known." Key features might be labeled *Je suppose,* meaning "I suppose." These terms have been abandoned, but in reality they could be applied to vast stretches of the deep sea.

Even shallow regions of the ocean warrant labels of *nondum cognita* and *Je suppose.* The basic geography of the depths stubbornly resists inventory. Deep and shallow waters, what sailors and submariners sometimes call thick and thin waters, remain poorly mapped in much of the world. This is not a matter of navigation charts missing irrelevant minutiae. This is a matter, in places, of missing mountains. And while a peak that does not reach close to the surface might be irrelevant to, say, a tanker skipper, it can be a matter of utmost significance to someone like Commander Chapman and his colleagues.

It remains possible for a highly sophisticated, expensive, well-crewed American military submarine to discover new mountains in the worst possible way—that is, nose-first. This possibility is not hyperbole but inconvenient fact.

On January 8, 2005, the Los Angeles class USS *San Francisco* fast attack nuclear submarine traveling at a depth of 525 feet ran head-on into a seamount, an underwater mountain that did not exist on her captain's charts. The crash injured ninety-eight crewmen. One man died. The submarine's pressure hull—the part of the submarine in which the crew worked and ate and slept and dreamt of home—remained intact, but the forward ballast tanks were ruptured.

Submarines use ballast tanks to provide buoyancy, to float, to return to the surface. Because military submarines do not dive to truly extreme depths, they can rely on pressurized gas to fill ballast tanks, displacing water to provide flotation. The tanks are not part of the submarine's pressure hull, but they are nevertheless critical to survival. A ruptured ballast tank in a submarine is comparable to a yawning hole through the underside of a surface ship.

The crew scrambled to save the *San Francisco,* and with her their own lives. Injured and uninjured put their training to the test. The classes, the drills, the simulations, and the collective experience of all aboard were called upon to keep their boat from spiraling into her last dive, from submerging to crush depth and near-instantaneous flooding with inescapable death.

Where her bow had been, there was now dented, twisted, and sheared-off metal, a wound so gaping and profound that the vessel's survival, her ability to emerge from beneath the waves and stagger back to port, is nothing less than astounding. After the crash, surfacing was not a certainty for the *San Francisco.* But surface she did.

The accident happened a few hundred miles from Guam—not far, as ocean distances go, from the Challenger Deep and the location of the *Trieste* dive forty-five years earlier.

The damaged bow of the USS San Francisco, in dry dock after she struck an under-water mountain in 2005. (U.S. Navy, Photographer's Mate 2nd Class Mark Allen Leonesio. "The appearance of U.S. Department of Defense [DoD] visual information does not imply or constitute DoD endorsement.")

Again as fact rather than hyperbole, it is true to say that the geographies of both Venus and Mars have been more accurately mapped than the geography of Earth, thanks to the oceans that currently blanket two-thirds of the planet. Radar and other space-based tools can peer through the thin atmosphere of Mars and even the thick acid fog of Venus, but they cannot penetrate the depths of the sea. It would not be possible for a careful navigator equipped with the best available charts to inadvertently run into an unknown mountain on Venus or Mars, but for submariners, for people like Commander Chapman and his colleagues, the potential remains for the unfortunate discovery of unexpected and unwelcome peaks.

In my conversation with Commander Chapman, I did not ask his opinion about the *San Francisco* accident. That, it seemed to me, would have been indelicate. I did, however, ask if he knew any good submarine jokes, just as I had aboard the *Tennessee*. He, too, had nothing to offer, or so he said. If, in fact, he knew any submarine jokes, they were either classified or perhaps too crude to share during an interview, or maybe they were outright embarrassing groaners not worthy of repetition without the presence of copious amounts of alcohol. This is disappointing, because I like jokes. And it seems to me that comic relief should be an important component of any topic as serious as military submarines.

With a bit of research, I turned up a few examples that may explain Commander Chapman's silence.

A one-liner, a riddle more than a joke: "What's long and hard and full of seamen?"

And this one, less crude: "Did you hear about the sub that managed to get a couple of inches of water in the bottom? They drilled a hole in the floor to let the water out."

And one more: "A submarine had just set a record for number of days submerged. The commander and his crew had been underwater for several long months, traveling up the Atlantic and under the Arctic ice pack and down through the Pacific and around Cape Horn before returning to their base. A reporter asked the commander why he had stayed down for so long. The commander thought about this question for a moment and then responded, 'I didn't want to come up before my submarine did.'"

⌒

"Enemy submarines are to be called U-boats," Winston Churchill purportedly once said. "The term 'submarine' is to be reserved for Allied underwater vessels. U-boats are those dastardly villains who sink our ships, while submarines are those gallant and noble craft which sink theirs." He was not joking.

The good people of U-Boat Worx in the Netherlands may never have heard Churchill's words, or if they have, they may not have paid attention. They may not have cared about the words of a dead British politician.

The submarines of U-Boat Worx were never intended to sink any vessel, theirs or ours. And they are not, by most standard definitions, submarines at all. They are submersibles. Submarines can move long distances on their own, either on the surface or at depth. Submersibles can go reasonably deep, but they lack the ability to travel long distances. Moving a submersible from one location to another relies entirely on a surface support ship. In the case of U-Boat Worx, the surface support ships are typically megayachts belonging to ultrahigh-net-worth individuals. The annual operating cost for the support yacht might exceed the two- or three-million-dollar price tag of the submersible.

U-Boat Worx makes the submersibles featured in the book that I left with Commander Chapman and his crew. When I call the company's sales department to request a price list, I am told that a deep-diving personal submersible is not the sort of thing that one should choose based on price. Bargain hunting and diving to thousands of feet do not mix, not even in the Netherlands, a nation known for frugality.

If I have ever regretted my utter failure to reach billionaire status, it is when I realize that I will never own a submersible capable of routinely diving beyond the reach of sunlight. But still, I am drawn to both the idea and the reality of a personal submersible, a way to explore great depths. There can be no harm in window-shopping.

And so I make my way to Breda, not far from Amsterdam and a mere ninety miles from the birthplace of Cornelis Drebbel, the Dutchman whose invention plied the Thames and was, by modern standards, more of a submersible than a submarine.

Breda is a city of about 300,000 people. I find construction sites, a sign describing the World War II bombing of a church where a small office building now stands, suburban housing, pubs, real estate offices, a car dealership. I wander for some time in the rain, looking for the home of U-Boat Worx. When I ask passersby for directions, specifying that I am looking for the place that builds little submarines, they look at me with Dutch eyebrows raised. The Netherlands, once but no longer one of the world's greatest sea powers, is not known for building submarines. And Breda is not a maritime city. Several times, my question is answered with another question. In one form or another, I am asked if I am certain that I am in the right city. "Breda? Are you sure?"

I know I am not only in the right city but in the right neighborhood, and close to the right spot. But where is U-Boat Worx? I walk past the same car dealership three times before I see the submersible manufacturer tucked away around a corner, behind the dealership, as discreet as a business building submersibles could possibly be. A simple logo says U-BOAT WORX, and through a window I see a partially assembled sample of their handiwork on the shop floor.

Submersibles are like atmospheric diving suits and submarines in that the cabin—the part that holds passengers and pilots—is not pressurized. A submersible's innards carry a pressure equal to or very close to that of sea level. Submersibles descend as deep as the strength of the vessel's hull and hardware allow. Inside, people breathe normally. The cabin atmosphere is pulled through a chemical filter that removes exhaled carbon dioxide. Oxygen is injected as needed. Passengers and pilots descend and return to the surface without any of the concerns that plague divers exposed to the pressures of deep water. There is no requirement for decompression, and there are none of the toxicity problems that come with breathing air or other gases at great depths. Submersibles control their depth with ballast tanks and thrusters, just like full-size submarines do, but without the range and firepower.

Basic submersibles, it turns out, are surprisingly easy to build. Hobbyists occasionally construct submersibles. With materials available from local stores, a high school student can put together a shallow-diving submersible. Although not an entirely safe approach, hulls can be fashioned from plastic sewer pipe, and thrusters can be assembled from electric trolling motors driven by car batteries. An American named Karl Stanley, trained in history rather than engineering, made his own, increasingly well-known *Idabel,* capable of carrying him and two paying passengers to depths beyond two thousand feet.

So with modern materials and knowledge, it is not so difficult to build a submersible. But it is difficult to build one that is safe, that can dive deep, and that can connect to a market best characterized as existing in a state of rapid evolution.

Inside U-Boat Worx, Roy de Boer, the company's communications and marketing executive, walks me through the shop. Several submersibles are in different stages of construction. Completed models look not at all streamlined but at the same time enchanting. Had Jules Verne and Jacques Cousteau shared a bottle of wine and a sketch pad, the resultant drawing might have looked like a U-Boat Worx submersible. Three words come to mind: "underwater functional techno-art."

The company makes nine models. Each one has a unique appearance but is immediately recognizable as a machine that belongs underwater. Here on the shop floor, on the day of my visit, a model the company calls the C-Explorer 5 is little more than a bare hull—a transparent acrylic dome mated to an acrylic cylinder mated to a steel dome. In a month, it will be ready to dive. The now empty hull will hold stitched leather seats, a touch screen computer system, a high-end sound system, and controls that would feel familiar to any experienced gamer. Outside, the surrounding frame will hold thrusters, battery banks, spotlights, and oxygen tanks. Owners can choose various additional options, such as a manipulator arm.

Some submersibles, including most of the world's deepest-diving research submersibles, require observers to peer through small, round portholes. The acrylic hulls of luxury submersibles give their passengers and pilots a panoramic view and what is best described as an immersive experience. On the surface, the hull is transparent but visible, like thick curved glass. Underwater, the hull, which has light-transmission characteristics similar to those of seawater, becomes all but invisible. If it were not for the dry clothes and comfortable surroundings, passengers and pilots might think that nothing separates them from the sea itself.

Like many of the parts used in U-Boat Worx submersibles, the acrylic domes—which can be almost ten inches thick—are built by a third-party manufacturer. They come from a British-based company that makes similar acrylic products for everything from hospital hyperbaric chambers to commercial aquariums. The touch screens are made by computer manufacturers. The thrusters are sourced from underwater robotics companies. The batteries, of course, are built by battery manufacturers.

Most of the pieces are not custom-made, but these submersibles are by no means spit out by anything resembling an assembly line. They come together through what might be described as modular construction. At one bench, two technicians test parts. Nearby, a pressure hull takes form. Around a corner, other technicians wire batteries. Elsewhere, interior finishing touches take shape within the confines of another pressure hull, a task reminiscent of the construction of a ship in a bottle.

The company began manufacturing submersibles in 2005 and has, as of the day of my visit, delivered twenty-two of their machines. The roughly thirty full-time employees include around ten engineers, and together they build five to seven submersibles each year, at least one of which is likely to be a new model.

Like its competitors, U-Boat Worx has to respond to the needs of clients. There was a time when the market seemed interested in submersibles capable of holding no more than an owner and a passenger. Soon it became clear that few billionaires wanted to pilot their own machines. Also, they wanted to bring guests along. Two seats were inadequate. Three-person submersibles were needed to provide space for a pilot, the owner, and at least one guest. Now, one of the U-Boat Worx models can carry nine passengers. It is quite possible to host a small party hundreds of feet below the surface.

There is, too, the issue of launch and recovery. Owners, in most cases, use their submersibles from megayachts that were not originally built for submersible support. But still, they want something that can be launched and recovered with relative ease. Naturally, they demand comfort. They do not want passengers aboard the submersible as it is swung over the side. The submersible must float like a boat on the surface so that passengers can board and disembark. But owners certainly do not want anything blocking the view underwater, so a large hull beneath the passengers is out of the question. One solution: the addition of inflatable tubes that provide buoyancy and stability on the surface, similar in appearance to those of an inflatable boat. Passengers board from a launch as the submersible floats on the surface. Upon descent, the tubes disappear, deflating under the pressure of the sea. After surfacing, they are reinflated. A launch picks up passengers before the submersible is lifted back aboard the owner's megayacht.

And through all of that—from boarding to descending to ascending to disembarking—there is the issue of safety. Where owners, guests, and pilots are headed, to depths of 3,000 feet, pressures outside approach those found inside a half-full scuba tank. The pressure hulls that protect passengers are tested before they are shipped to Breda. After assembly, the finished product is tested in a tank that sits behind the U-Boat Worx shop. A third predelivery test takes place in Holland's Grevelingen estuary. And, finally, there is a sea trial at a location of the new owner's choice. Before it is over, the entire submersible is subjected to a depth half again as great as its maximum rated diving depth—a submersible intended to dive to 3,000 feet, for example, will be subjected to 4,500 feet. And along the way, each submersible is certified by the independent international certification organization DNV GL.

But what if, for example, a submersible becomes entangled at depth, as happened to the research submersible *Johnson-Sea-Link* in 1973, resulting in two fatalities? The parts of the modern luxury submersibles most likely to become entangled, such as manipulator arms and thrusters, can be dropped, allowing the rest of the submersible, its pilot, and its passengers to ascend.

But still, what if something goes terribly wrong? Failure of the pressure hull is extremely unlikely, but what if a submersible were trapped on the seabed, despite its ability to cast off exterior parts?

U-Boat Worx sells more than just submersibles. They also sell operational expertise, maintenance plans, and services. They offer advice on where to dive, on where to find interesting deep reefs, lantern fish, stalked crinoids, goblin sharks, and deep wrecks in clear water. They can help, too, with permits, a requirement in many nations troubled by terrorism, espionage, and smuggling. And they provide safety plans. When a U-Boat Worx submersible dives, it is alone on the bottom but tracked from above. The pilot communicates with topside staff. If the worst were to occur, the emergency plan would be put into action. That plan would probably call for the immediate release of a buoy. It might call for the assistance of another, nearby submersible or an underwater robot. In shallower water, the plan might involve technical divers, trained and equipped for short dives to hundreds of feet.

In the worst case, a submersible trapped on the bottom carries at least ninety-six hours of oxygen for everyone onboard. The passengers might run out of champagne, but they are unlikely to suffocate before responders bring them back to the surface.

⌐

Modern small submersibles, like those built by U-Boat Worx, were not invented with the luxury market in mind. There was a time when they were associated with the research community and, later, with offshore industries that laid cables and pipelines or worked in oil fields. In both cases, the market more or less dried up. The submersibles worked well enough and could be, by themselves, cost-effective. But the support vessels were another story. A submersible needed a ship, and that also meant a full staff had to be on location to support the work of one or two people down below.

Over time, submersibles became increasingly scarce. Despite today's luxury market, it is fair to say that there are more retired submersibles than there are active submersibles. But a few of the older submersibles that remain active deserve to be better known, to be celebrated.

There is, for example, *Alvin*, launched in 1964 and owned by the Woods Hole Oceanographic Institution. Don Walsh himself provided input into its design, although few seem to remember his role. The little submersible, capable in her current configuration of carrying three people to a depth of 20,000 feet, has been attacked by a swordfish, located a lost hydrogen bomb, and explored the *Titanic*. She, or the people aboard her, discovered the hydrothermal vents known as black smokers that fuel Phil

Nuytten's dreams of Vent Base Alpha. And she has sunk, been salvaged from a depth of 4,900 feet, and been repaired and recommissioned. She is the subject of the wonderful 1990 book by Victoria A. Kaharl, a biography of sorts, called *Water Baby: The Story of* Alvin.

The submersible Alvin *in 1978.* (National Oceanic and Atmospheric Administration, Department of Commerce, and OAR/National Undersea Research Program [NURP]; Woods Hole Oceanographic Institute.)

There are also the Russian submersibles *Mir 1* and *Mir 2*, which, like *Alvin,* are capable of dives to 20,000 feet. And also like *Alvin,* they have visited the *Titanic.* In addition, they have been to the geographic North Pole, reaching the seabed at 14,000 feet and stirring up both interest and controversy by leaving behind a Russian flag made of titanium. Among the passengers who have submerged in the *Mir* submersibles are Canadian filmmaker and explorer James Cameron and Russian leader Vladimir Putin.

The French have the *Nautile,* capable of dives to 20,000 feet. The Japanese have the *Shinkai 6500,* engineered to go to 21,000 feet. And the Chinese have the *Jiaolong,* meaning something like "water dragon" or "flood dragon," which can roam to 23,000 feet.

The world's submersible fleet puts ninety-eight percent of the seabed within reach. But the deeper depths, like those enjoyed by Don Walsh in *Trieste,* cannot be reached by any submersible existing as I write these words.

Submarines are busy on missions of military value. Luxury submersibles sit aboard megayachts owned by individuals of ultrahigh net worth, whom, in all probability, I will never meet. Research submersibles are scarce and fully booked by researchers. No one, in all of my pesky telephone calls and interviews and site tours, has offered me a ride. These undersea vehicles, it seems, do not pick up hitchhikers. And so I turn to Karl Stanley and his *Idabel,* on the island of Roatán, off the coast of Honduras. Karl will, for a fee, take me to 1,500 feet. For that matter, he will take anyone, for a fee, to 1,500 feet, or deeper.

I mention research and luxury submersibles to Karl. "You know," he says, "the deep sea should not be the private playground of government scientists and the über-rich. It belongs to all of us." With that philosophy in mind, and without credentials in engineering or science, and without institutional support, he built *Idabel*. It is thirteen feet long and seven feet tall, not unexpectedly painted yellow, and it has been to 2,660 feet.

As with most great opportunities, diving in *Idabel* comes with catches. First, it is not an inexpensive ride. Second, neither Karl nor *Idabel* has been certified as safe by independent authorities. This does not mean they are dangerous, but it does mean putting a certain level of trust in a stranger and his homemade machine.

Idabel sits in a cradle on a dock in the Roatán tourist ghetto of Half Moon Bay. One of the beauties of Karl's operation is that he is little more than a stone's throw from deep water. He does not need the inconvenience of a support ship. He operates directly from his dock.

My wife, who after all is a marine biologist, shares my fascination with the underwater world. We board *Idabel* through her single hatch and move forward to sit on a narrow bench in the passenger compartment.

In front of us is an acrylic dome thirty inches in diameter. At our feet is a flat porthole, thirteen inches across, looking downward.

Karl climbs in behind us, in the pilot's space. He lowers the vessel's cradle into the water. He casts off, and for a short while we feel the light surface chop of Half Moon Bay rocking *Idabel*. Sun shines upon her yellow hull as we maneuver toward the reef cut.

The annoying rumble of internal combustion comes to us through our still open hatch. It comes from the engines of glass-bottom boats, of boats carrying snorkelers, of boats loaded with scuba divers. We take comfort in the steady, unobtrusive hum of *Idabel*'s own electric motors.

Behind us, above the waves, Roatán rises steeply upward into tropical skies. In front of us, beneath the waves and beyond the reef cut, the shallow seabed plunges downward. The island is best described as the upper extremity of a mountain range rooted in the Cayman Trough. A geologist might say that Roatán stands as an emergent expression of the mostly drowned Bonacca Ridge.

Within a few miles of *Idabel*'s current location, the seabed lies beneath ten thousand feet of water. Between here and there, the bottom is a mix of steep walls and furrows, exposed basalt and limestone, jumbled rocks and sand, living coral and dead silt.

Minutes from the dock, close to the reef cut, Karl pulls the heavy steel entry hatch downward. With a dull thud, he seals the three of us into *Idabel*. The outside noise, aside from the hum of our own motors, is locked out. Karl's view of the world now comes through the nine portholes that surround him, and ours comes from the dome in front of our noses and the porthole at our feet.

In the relative quiet, an interior fan hums audibly. The fan is not designed to offer a cooling breeze, but rather to suck air through the carbon dioxide scrubber. Over the next few hours, Karl will occasionally bleed pure oxygen into our tiny shared atmosphere. It is, in principle, the same familiar system used in old-fashioned submarines, atmospheric diving suits, and rebreathers.

Karl turns a valve. A bubbling gurgle—air bleeding away from ballast tanks—overwhelms the electric hum. *Idabel* dives, and we dive with her.

At 10 feet, the rocking of the surface chop subsides. Through our acrylic dome, we see the ocean around us, majestically blue and clear. At a depth of 30 feet, we look upward. We see a glass-bottom boat directly overhead. We cannot quite make out the faces of tourists behind the glass, but we know they are there. We know they stare as we disappear beneath them.

We pass through the reef cut, a narrow gorge with living walls of coral. We are in the open ocean now, on the side of the submerged mountain that is Roatán. We descend down a nearly vertical wall.

At 80 feet, we pass a group of scuba divers. They point at us, as if they have stumbled upon a leatherback turtle or a whale shark or a manta ray. We wave.

We pass 132 feet, the deepest I have been on a breath of air.

We pass 525 feet, the depth at which the USS *San Francisco* unexpectedly discovered an undersea mountain of her own.

At 700 feet, the vertical wall intercepts a sloping boulder field. Perspective presents a challenge, but this is no ordinary boulder field. This is neither glacial moraine nor blasted rock. In the dark blue twilight penetrated by *Idabel*'s lights, broken pieces of the wall that we have just descended stand high above the seabed. They are giant chunks of subsea cliff that have landed here. Some are ninety feet tall.

Through our acrylic dome, we see stalked crinoids, relatives of the feather stars familiar to snorkelers and scuba divers, but beneath their pinnate tentacles, long stems—which their shallow-water cousins do not have—attach them to the rock wall. Illuminated by *Idabel*'s lights, the stalked crinoids might be likened to miniature bright orange palm trees growing from a towering cliff. Creatures of their sort went extinct long ago in shallower water.

At a depth of 1,300 feet, Karl flips a switch, turning off *Idabel*'s lights. Ahead, the shadows challenge vision. Above, as we look straight up, light struggles, penetrating as a faint indigo glow. "This is as deep as I have navigated without my lights," Karl says, much as someone else might say, "I carry a flashlight on my evening walks."

Karl flips the switch back to the On position. *Idabel* lights up the depths. She descends. Her electric motors whir.

The black needle on the pneumofathometer—the depth gauge—reads 1,500 feet. We have passed the likely crush depth for many military submarines.

We level off. We have paid for a dive to 1,500 feet. Deeper dives cost more. Our current depth has put enough pressure on our budget. We will go no deeper.

Karl once again extinguishes the lights. Above, as we look

straight up, darkness reigns. Likewise, as we look forward, down, port, and starboard, all is lightless. It is half past midnight in the afternoon. This is no place for a nervous disposition.

And yet what a place it is. It is not just what we see outside, but simply the gestalt of being here, of being alive and warm and happy 1,500 feet beneath the sea. It is the knowledge that a solid hull keeps something like 750 pounds per square inch of water pressure from entering our little space, pressure that, cumulatively, applies a weight of 15 million pounds to *Idabel's* hull.

And yet, we have barely penetrated the uppermost layer of the deep sea. For submersibles like *Alvin* and *Shinkai 6500* and *Jiaolong,* this would be shallow. For *Trieste,* this depth would be very near the beginning or end of a dive. For someone like Don Walsh, my feelings of delight, my use of the word "gestalt," and my willingness to pay more than I can afford to be here might all seem a little laughable.

And my thoughts about Don remind me of his view of the future. The heavy lifting of deep sea exploration, he predicted, would be done not by submersibles but by robots.

Here at 1,500 feet, I try to imagine a little robot swimming about, all alone, headed deeper, downward into the welcoming blackness. And I smile.

Chapter 6

THE ROBOTS

Robots, both above water and underwater, real and imaginary, have been around for hundreds of years. In 1206, a Middle Easterner named Ibn al-Razzaz al-Jazari described various mechanical robots in *The Book of Knowledge of Ingenious Mechanical Devices,* including what amounted to a robotic washroom attendant and a floating robotic quartet. Near the end of the fifteenth century, Leonardo da Vinci designed and possibly built a knight in armor capable of sitting, waving his arms, and moving his head. By the eighteenth century, the Japanese were making tea-serving robots. And the industrial revolution was industrious in part because of manufacturing systems that were in everything but name factory robots, machines capable, for better or for worse, of doing work once done by humans. And if these mechanical devices were early robots, then it is but a small stretch to say that the simple and humble sampling dredge might be seen as an early underwater robot, a mechanical extension of humanity that reached beneath the waves to bring something back.

In 1815, almost a century and a half before *Trieste,* Edward Forbes, dredger extraordinaire, was born on the Isle of Man in the Irish Sea. At seven years old, he was collecting curiosities from beaches and mudflats. At twelve, he wrote *A Manual of British*

Natural History in All Its Departments. At twenty, he published a paper on dredging in the Irish Sea, sharing descriptions of what had come to the surface in simple dredges.

A dredge similar to those used by Edward Forbes, and men sorting the contents of a dredge on deck. In use, the dredge would be pulled across the bottom, scooping up mud, animals, and plants. Mud and small objects would pass through the mesh of the dredge, but larger plants and animals, caught in the mesh, could be hauled to the surface. Based on samples collected in the Aegean Sea, Forbes suggested "zero of animal life possibly about 300 fathoms." (Left: *Encyclopedia Britannica*, 1911, Volume 8, Edition 11) (Right: National Oceanic and Atmospheric Administration, Department of Commerce, Archival Photography by Steve Nicklas [NOS,NGS])

Forbes used small sampling dredges, little more than nets attached to heavy frames that scratched along the seabed as they were pulled by cables trailing behind small boats. He can be imagined laboring at his oars, holding tension on one of his dredge cables, counting on waves to add further bursts of tension as his little skiff rode the crests. Down below, in the twilight murk, the dredge made jerking progress as it bit through the seabed, sending up clouds of silt. And after a time, Forbes brought everything to the surface and dumped his catch at his feet, where he sorted through mud and sand and perhaps old boots to find snails, clams, fish, and sea cucumbers.

In an era when collecting and displaying natural curiosities were all the rage, people came to young Forbes seeking advice. He sketched a comical figure of a dredge in action, and he wrote a dredging song:

> Hurrah for the dredge with its iron edge,
> And its mystical triangle,
> And its hided net with meshes set,
> Odd fishes to entangle!
> The ship may rove through the waves above,
> 'Mid scenes exciting wonder;
> But braver sights the dredge delights
> As it roveth the waters under!
> Then a dredging we will go, wise boys!
> Then a dredging we will go.

Although he was not the first to explore the depths with a dredge, his work led to the establishment of a dredging commit-tee at the British Association for the Advancement of Science in 1839. Two years later, it also landed him a berth aboard HMS *Beacon*, bound for the Aegean Sea, where he would spend a year and a half sampling and drawing the biological treasures that came to the surface and thinking about what it all meant. He had graduated from rowboats and sailing skiffs to a ship with a crew. He now supervised sailors who did the heavy work with the dredge cables.

In the Aegean, his dredge reached a depth of 1,380 feet. To his mind, at a time before the true depths of the seas were understood, he was bringing up material from the abyss itself.

His sixty-five-page report on the expedition, published in 1844, is a tedious collection of tables punctuated by dull text. It contains the names of many species, one after another. But the report also divides the bottom of the Aegean into eight regions. The deepest, the eighth, starts at 630 feet.

Forbes observed, "The number of species and of individuals diminishes as we descend, pointing to a zero in the distribution of animal life as yet unvisited." At the bottom of a long table he called "Diagram of Regions of Depth in the Aegean Sea," he wrote two concise sentences: "Zero of animal life possibly about 300 fathoms. Mud without organisms remains." In other words, the deeper he went, the less he caught.

Edward Forbes offered a fanciful sketch of a sampling dredge in action as the frontispiece of his posthumously published 1859 book, Natural History of the European Seas. *The sketch suggests that Forbes realized that fish and jellyfish routinely avoided his dredges.* (National Oceanic and Atmospheric Administration, Department of Commerce, Naval History and Heritage Command, and Steve Nicklas [NOS, NGS])

Although he did not dwell on it, the deeper he went, the more difficult the work became. The cable to the bottom got longer and longer. The time required to lower and raise equipment increased. The effort of pulling something through so much water from a moving boat taxed the muscles and the mind. The sailors assigned to the backbreaking labor grumbled. Deepwater dredging was physically hard. And for what? Piles of mud and, for the most part, inedible oddities.

The fact that Forbes never reached a depth at which his nets reliably came up empty, a depth at which he could absolutely count on catching nothing at all, did not prevent him from speculating on what became known as the azoic zone, the region of zero animal life. He caught less as he went deeper, and so, he thought, if he continued downward, he would catch nothing at all, because, he guessed, there was nothing there to catch. Later, he elaborated on the abyss as a place "where life is either extinguished, or exhibits but a few sparks to mark its lingering presence."

Forbes, even at a young age, was a well-respected scientist. He traded correspondence with the likes of Charles Darwin, discussing seashells and barnacles. He published works on terrestrial plants as well as marine life. He became a professor of botany at King's College London, served as a paleontologist with the British Geological Survey, and was eventually awarded a chair at the University of Edinburgh.

Over time, Forbes's flawed speculation about the azoic zone, about the lifelessness of the deep, took on the patina of fact. People wanted to believe in a lifeless abyss and could not imagine animals in the constant darkness, the chilling cold, the incredible pressure. The deep sea environment was far too alien for life. So much so that samples of organisms recovered from great depths were duly ignored. For example, living organisms on oceanic telegraph cables hauled up for repairs were not accepted as evidence of life at depth. To anyone predisposed to believing in a lifeless deep sea, it was obvious that the organisms attached themselves to the cables during lowering and raising, in mid-water. Never mind that the organisms looked more suited to crawling on the bottom than to swimming. Forbes's suggestion, his vision of lifeless depths, his azoic hypothesis, made perfect sense to those adapted to life in air and restricted to the shallowest of shallows.

Forbes died before his fortieth birthday. He does not seem to have vigorously defended his speculation about "zero of animal life" or to have thought of it as an important contribution. And yet, to this day, he is remembered in textbooks and lectures as the man who spawned the persistent vision of a sterile deep sea, the man who was dead wrong about what one might find while poking around down there. Sadly and without real justification, he is remembered by many as something of a deep sea buffoon.

When I read about his life, when I read his reports, I think that it would be fun to have some sort of mechanical extension of myself that could reach beneath the waves to bring something back. I envision myself lowering it over the side of the boat that my wife and I call home and pulling it along. But I recognize that times have changed, that my wife and I may not want all that mud on the deck, and that I do not want to cut a swath of destruction across the seabed.

I realize that what I need is not a dredge, but a twenty-first-century underwater robot. I may not be able to afford my own submersible capable of diving deeper than I can go on a breath of air, farther down than I can conveniently go breathing compressed helium and oxygen, but perhaps I can buy a robot that can go there for me.

⌐

In addition to dredges, there were mechanical devices for sampling water, grabbing sediment off the bottom, and clearing mines and debris. All of them could be, in a pinch, thought of as early underwater robots, or at least as precursors. There were also cameras lowered over the sides of boats and sometimes towed, including, for example, one that reached 4,200 feet in 1939.

But credit for the first modern underwater robot usually goes to the French explorer and archaeologist Dimitri Rebikoff. Sometimes working with Cousteau, he invented various kinds of camera gear for divers. He also built and used what today would be called an underwater scooter, a contraption for pulling a diver behind what looked like a stubby torpedo with a battery-powered propeller. Melding these technologies, and then adding control wires that reached all the way to the surface, he invented the robot he called Poodle in 1953. An operator saw the seabed through Poodle's camera and controlled her location with her propellers. During her first day of trials, the robot—well, her operators—found two Phoenician shipwrecks, one at 500 feet and the other at 740 feet.

Eight years later, in 1961, the U.S. Navy took delivery of what would evolve into its Cable-Controlled Undersea Recovery Vehicle, or CURV, initially used to recover test torpedoes. It was as big as a small submersible, about the size of a minivan. It was equipped with cameras and thrusters, buoyancy chambers and sonar. Its claw—its manipulator arm—was large enough to salvage torpedoes off the seabed.

In 1966, CURV I stepped away from its day job long enough to recover the intact hydrogen bomb that had been lost in 2,900 feet of water off the coast of Spain. The submersible *Alvin* had found the bomb but failed to recover it. CURV I brought the bomb to the surface.

The first CURV spawned a second and third model. In 1973, CURV III rescued two men trapped in a submersible off Ireland in 1,575 feet of water. Echoing the hydrogen bomb incident, attempts to recover the submersible using another submersible had failed before CURV III succeeded. By then, the two men in the submersible had stretched their seventy-two-hour emergency life support systems to seventy-five hours. One of the men rescued wrote a book about the experience, appropriately titled *No Time on Our Side*.

In both cases, the CURVs had succeeded where divers could not go and manned submersibles had failed. The CURVs relied on what was, for the time, sophisticated technology, but as important or even more so, they succeeded because they could be sacrificed. They were expensive, but they were expendable. No humans were aboard.

CURV II, an early remotely operated vehicle (ROV), or underwater robot, developed in the 1960s by the U.S. Navy and its contractors to retrieve test torpedoes. (U.S. Navy)

In the hydrogen bomb recovery, CURV I became hopelessly entangled in the bomb's parachute, after which both were hauled in using the robot's tether. In saving the trapped submersible pilots, CURV III attached a line to the submersible, but when the robot's tether became entangled with the line, jeopardizing the rescue mission, CURV III was unceremoniously cut loose and abandoned.

"CURV," proclaims a low voice in a propaganda film from the 1960s, "has extended man's hand deep into the sea." And that hand was quickly adopted by oceanographers and even commercial entities, including, for example, investors interested in harvesting manganese and phosphate nodules off the coast of California.

Underwater robots became known, generically, as remotely operated vehicles, or ROVs. The early ROVs, despite their usefulness, were even more clunky than, say, atmospheric diving suits. A crane was needed for launch and recovery. Underwater, housings intended to keep the electronics dry leaked, as did hydraulic fittings. Thrusters were noisy enough to interfere with onboard sonars, and, worse, they were unreliable. Early ROV cameras, intended for the low-light world of the deep sea, could be damaged by sunlight. Everything required constant maintenance, and even when working properly, early ROVs were hard to control. They were, in other words, well positioned for improvement. And improved they were, first by the military and later by the same industry that took saturation diving from the U.S. Navy and put it to work: the offshore oil and gas industry.

On August 27, 1859, well removed from the nearest ocean, a well drilled to a depth of just less than seventy feet underground struck oil near the backwater logging community of Titusville, in northwestern Pennsylvania. This was by no means humanity's first encounter with what was then known as rock oil, to distinguish it from oil that came from whales and other sources, but it was the first time that rock oil became available in large quantities. And it was this discovery that triggered a boom and launched what since has been called an addiction.

That year, two thousand barrels of rock oil were taken from the ground. Ten years later, more than four million barrels were extracted in America alone. Today, more than eighty million barrels of oil—more than three billion gallons—come up every single day.

Before the end of 1859, the Pennsylvania oil boom had spilled over into neighboring Ohio, and by 1865 it had traveled all the way to California. Around 1891, drillers realized that they could access oil under Ohio's shallow Grand Lake St. Marys from wooden platforms built over the water itself. By 1896, California drillers were working from piers extending into the Santa Barbara Channel. Within twenty-five years, there were wells in Azerbaijan's Caspian Sea, in Lake Erie, in Louisiana's Caddo Lake, in Venezuela's Lake Maracaibo, and in the coastal waters of the Gulf of Mexico. By the end of 1947, the Louisiana oil fields had moved out of sight of land. And by the end of the 1960s, the North Sea became yet another important source of oil. At the turn of the twenty-first century, offshore oil fields operated in India and the Persian Gulf, off Angola and Brazil, in the Barents Sea, the Beaufort Sea, and the Kara Sea. And elsewhere.

As shallow-water oil reservoirs were exhausted and oil prices rose, drillers moved to deeper water. As early as 1977, Shell Oil

This photograph of oil wells on piers extending off the coast of Santa Barbara County, California, was taken by G. H. Eldridge sometime before 1903. The development of offshore oil and gas fields necessitated significant progress in underwater technologies. (National Oceanic and Atmospheric Administration, Department of Commerce, and Pacific Committee of the American Association for the Advancement of Science)

began constructing its Cognac platform in 1,025 feet of water in the Gulf of Mexico. In 2007, BP brought oil up from its Atlantis field, with a depth of 7,100 feet. And in 2017, Shell extracted oil from a depth of 9,500 feet in the Gulf of Mexico. At 9,500 feet, the water pressure is something like 4,700 pounds per square inch.

In the earliest days, the water was too shallow to require divers. Later, divers wearing the heavy copper helmets and rubberized canvas suits that would have been familiar to Augustus Siebe found employment working on pipelines and under platforms. As depths increased, heliox and atmospheric diving suits and the saturation diving techniques developed by George Bond and others came into play. And diving became more expensive. There were the expenses of paying the diver and the support crew, of renting the equipment needed to support the dive, and of buying helium at costs exceeding one dollar per breath. But there was also the cost of the ship, barge, or drill rig, of tying up fifty or a hundred workers while waiting for one person—the diver on the bottom—to complete some mundane but critical task, such as hooking up a shackle or tightening a clamp.

And there was the issue of safety. Divers suffered from decompression sickness. Working alone in the dark and the cold while breathing exotic gas mixtures, they lost their fingers in pipe flanges, their arms and legs to unseen underwater swinging crane loads, their vision to faceplates smashed in by the small explosions that sometimes came with the use of underwater cutting equipment. All too often, oil field divers simply died. There was a time when injuries and fatalities were an accepted part of any construction project, especially underwater, but even then, hurt and dead divers added to project costs. And, slowly, society changed its views regarding industrial injuries and fatalities. Oil companies realized that they could not blithely accept divers' injuries and deaths, even if those young men put themselves in harm's way of their own volition.

As the quest for oil drew operations deeper underwater, more than one engineer working in the 1970s and certainly in the 1980s realized that they were moving beyond the reach of divers.

While I am tracking down firsthand information about the oil industry's entrance into the world of underwater robotics, of ROVs, one name I hear repeatedly is Graham Hawkes. Now in his seventies, he has worked for decades on atmospheric diving suits and small submersibles, but also on ROVs. I find him, and in a rambling conversation he tells me about his work.

He once set a record for taking an atmospheric diving suit—a suit called the WASP—to a simulated depth of two thousand feet in a chamber. The chamber, he says, failed. That is, a seal or fitting in the chamber failed, sending it almost instantaneously to the surface. The WASP suit, with him in it, remained intact. Had he been breathing pressurized gas when the chamber failed, had he been a saturation diver, I would not be talking to him now.

He tells me of being sued by a large company claiming, incorrectly, that he had stolen their technology for articulated arms designed to work well under extreme pressure. He says that he nearly had a heart attack over the whole thing, but the case was eventually thrown out and the company that had brought the suit apologized and hired his services. He talks about working for Steve Fossett, a man who, in essence, collected records—records for ballooning, sailing, mountaineering, skiing, and flying. Had Fossett not died in an airplane accident in 2007, he might have claimed another record in a submersible designed at least in part by Hawkes.

But Hawkes also has this to say: "You have to ask, why not just use a remote? A pretty dubious argument favors the human senses in the water. Humans have five senses, but in a pressure vessel behind a five-inch-thick acrylic dome, they give up four of those five senses."

And he adds this: "The answer is in cost. People talk about risk, but really it's an issue of cost. Not only the cost of the vehicle but the cost of the support vessel."

And this: "The argument to put a human down is because 'I want to go there.'"

And, later, this: "No one should be able to say, 'I want to go, but the taxpayer should pay for my trip.'"

Regarding his ongoing work with manned submersibles, he says, "My clients in recent years have been extremely wealthy individuals." In other words, they are paying their own way, self-financing their own little adventure vacations far below the surface.

Going back to underwater robotics and the oil industry, Hawkes recalls a colloquium initiated by Sir Hermann Bondi, then the chief scientist at the British Department of Energy. The colloquium brought attention to diver safety, and Hawkes entered the discussion with the impression that submersibles and perhaps atmospheric diving suits would gain a boost, a shot in the arm. Instead, as Hawkes recalls and relates to me, the raised visibility went to ROVs.

Bondi was by training what most people would think of as a cosmologist and a mathematician, known in some circles as the cofounder of the steady state theory of the universe, but later in life he became involved with civil service in various influential capacities. In his autobiography, he wrote of his interest in ROVs: "One subject I was especially keen on was to promote the gradual replacement of divers in the North Sea oilfields by remotely controlled submersible vehicles to avoid having to use people in conditions of stress and danger."

He also wrote: "Remotely operated devices seem to me to be very much the technology of the future."

By 1974, only about twenty ROVs had been built. Of those, seventeen had been funded by various governments. But in the eight years between 1974 and 1982, something like five hundred ROVs were constructed, and as many as eight or nine out of ten were paid for by the private sector. By 1984, there were at least twenty-seven companies manufacturing ROVs.

Advances in miniaturization, materials science, and electronics paralleled the increased interest in ROVs. There were American, French, Canadian, Norwegian, Swedish, British, German, and Japanese ROVs, all taking advantage of new technologies. There were observation ROVs, little more than video cameras with thrusters that were often called flying eyeballs, and there were work-class ROVs, armed with manipulators and possessing the capacity to use tools. There were ROVs with names such as Spider, Sea Owl, Amphora, Cetus, Boctopus, and Scorpio. There were claims of stronger thrusters, better cameras, higher reliability, lower costs, and general superiority in all ways over anything offered by competing brands.

In the oil fields, a typical scope of work might involve inspecting a pipeline on the seabed or picking up a lost object. Many or perhaps even most projects, especially in the early 1980s, could bring more disappointment than satisfaction. It came in the form of ROVs that could not hold their location in a current, that became entangled on the legs of oil platforms or, even worse, in the propellers of support vessels, and that simply did not work because of failed lights, hydraulics, cameras, or sonar.

A later scope of work might involve more sophisticated jobs. An ROV, for example, might use a tool to cut through a two-inch-thick steel plate a thousand feet underwater. Or two thousand feet underwater. Or five thousand feet underwater. Or a group of ROVs might hover around a wellhead, all controlled by a team sitting far above, on the water's surface, working together through their various robots to solve some problem, to fix some bug in a realm entirely out of reach to the human hand.

And with all this complexity, sometimes with several ROVs swimming around a job site, trying not to get in one another's way, it is very possible that the entire project was worked out on simulators before anyone got in the water. The work had been practiced virtually, using the underwater equivalent of flight simulators, in a world where a programmer can control the effects of currents, visibility, equipment malfunctions, and maybe even human error, minimizing the time spent tying up expensive support vessels in very deep water.

While ROVs and all things mechanical will never be one hundred percent reliable, they grew in reliability to a point where operators could claim that their equipment was as dependable or

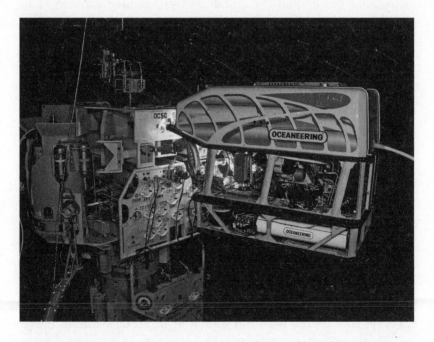

A modern ROV (at right) services an underwater valve assembly (at left) using a subsea torque tool. (This photograph was generously donated to the public domain by a vice president of Oceaneering, Duncan McLean)

more dependable than divers. Divers, after all, get hungry, cold, tired, and grumpy, and even the toughest come with annoying vulnerabilities, with an aptitude for injuries and an occasional susceptibility to mortality.

As ROVs became more sophisticated and drillers moved to deeper water, engineers realized that improvements in the robots themselves might not be the whole solution. The structures they placed on the seabed, the billion-dollar assemblies built in ship-yards in Louisiana, Scotland, and Korea, could be designed with ROVs in mind. For example, valve handles did not have to fit human hands, but instead could be built to be grasped by a mechanical manipulator.

All of this was happening somewhat quietly. It was not that the oil companies or their contractors were hiding what they were doing, but rather that the public showed limited interest. This was neither space nor a neck-and-neck Cold War race between the Soviet Union and the United States. This was not even Sealab. This was the oil industry working in the deep sea. It was industrial, dark, and boring.

And then, on the evening of April 20, 2010, at 9:45 local time, an explosion occurred on a drill ship contracted by BP. The ship was working in about 5,100 feet of water, nearly a mile deep. Eleven men were killed. Oil flowed into the Gulf of Mexico for four months, four weeks, and two days, until September 19. And, for much of that time, interested viewers—of whom there were many millions—tuned in to broadcasts coming straight from the lenses of ROVs working with relative ease at depths that not long before were unthinkable. Viewers could see ROVs using circular saws, diamond wire cutters, high-pressure water jets, and special cameras. It took the worst oil spill in history, but suddenly every-one knew something about ROVs.

And more than one viewer suddenly realized that a huge swath of the Gulf of Mexico had been industrialized. People like George

Bond and Phil Nuytten might talk of underwater colonies, of underwater farming and Vent Base Alpha, and I might think of nuclear submarines as underwater colonies, but the oil fields offer another kind of colony, for better or worse. The offshore oil fields will populate the depths, but with robots instead of humans.

None of this is to say that ROV development outside the oil fields ended. Take, for example, Japan's ten-ton Kaikō. She was launched in 1993. In her career, she found the wreckage of a ship sunk by an American submarine in World War II, discovered hydrothermal vents in the Indian Ocean, and connected cables for underwater measuring devices.

And Kaikō visited the Challenger Deep, making it the second vessel to go there, preceded only by *Trieste,* in 1960. She went there not once but many times, on several expeditions, swimming around at the whim of controllers who sat at the end of a control cable that was, by necessity, more than seven miles long.

Her surface support ship, *Kairei,* also deserves a few words. She was a built-for-purpose research vessel, 350 feet long. She was, like many of today's working vessels, dynamically positioned, meaning that she could pull up to a desired location — say, a point over the Challenger Deep — and stay put by virtue of computer-controlled thrusters. Anchoring over the Challenger Deep was not a convenient option, so dynamic positioning came in a little more than handy.

A hangar on deck stored Kaikō. To be launched, Kaikō moved aft along rails. Once she was in position near the stern, an A-frame lifted her off the ship and lowered all ten tons of her into the water. Her primary tether, about an inch and a half in diameter, ran off a spool mounted on *Kairei.* The spool was more than twenty-five feet in diameter. Kaikō descended at a rate of just over three feet per second while an automated tensioning system prevented the ship's motion on the waves from jerking at her tether. Three hours later, the massive spool was almost empty, and Kaikō was within a few hundred feet of this planet's deepest seabed.

For perspective, look out the window during a typical commercial airline flight. Find a space through the clouds to view the fields or ponds or forests below. The height of that flight aboveground is about the same as the depth of Kaikō. Now imagine that your flight cannot operate on its own, that it needs a cable attaching it to the ground. That was Kaikō's tether.

It would not be reasonable to expect an ROV to tow seven miles of cable through the water column. Kaikō was mostly a garage that just hung there, dangling beneath the ship. But another part of her swam away. The two parts were known, in English, as the launcher and the vehicle. The launcher—what I tend to think of as the garage—held another spool, a miniature version of the ship's huge spool, this one with about eight hundred feet of cable.

The vehicle, about the size of a car, used four horizontal thrusters and three vertical thrusters to move forward, aft, sideways, up, and down. The same thrusters allowed her to stand still, hovering. She had two onboard sonars, various oceanographic instruments, seven video cameras, a high-definition still camera, and two electrohydraulic manipulator arms. The manipulators moved with what robotics engineers refer to as seven degrees of freedom. The human arm, with its rotating shoulder joint and bendable elbow and swiveling wrist, has the same seven degrees of freedom. Kaikō's manipulators also provided what robotics engineers know as force feedback, meaning that the operator on the surface felt what the manipulator felt. If called upon to do so, Kaikō could pick up an egg without crushing it.

Among the things that Kaikō did not see on the Challenger

Deep seabed were fish. The fish reported by Don Walsh and Jacques Piccard could not be confirmed.

In fact, Kaikō saw very little in the way of life. Fans of Edward Forbes might smile at the news. His azoic hypothesis, they might think, could hold water after all, even if the depth was a bit off. But anyone believing that depth or its inevitable pressure limits life at the bottom of the Challenger Deep would be mistaken. The pressure affects protein stability and membrane permeability, but life has adapted to those inconveniences. Strange bacteria, some discovered by Kaikō herself, survive. A biologist might describe them as "obligately barophilic," because they can grow only under conditions of extreme pressure. If these bacteria were capable of perceiving anything at all, they would perceive that the shallows are inhospitable, foreign, and deadly. For them, depths above fourteen thousand feet are fatally shallow.

What else might limit life in the Challenger Deep? Food availability. The Challenger Deep sits under very clear water. Ordinary deep sea environments depend on dead plankton and fish raining down from above, but very little grows in the clear water above the Challenger Deep. What little food does fall down might be consumed in transit, by mid-water fish, squid, and filter-feeding plankton. In fact, it might be recycled several times in transit, with one beast eating another and a third eating the second, each stripping away calories. What remains for the inhabitants on the seabed would be nothing more than starvation rations. When Don Walsh and Jacques Piccard reported life at the bottom of the Challenger Deep, it appears, they were lucky to have hit a little pocket of activity, an oasis in a more or less foodless desert.

When I think of Kaikō, I find it easy to humanize her. I think of the Star Wars films, of the snooty C-3PO and the more rational but stubbier whistling R2-D2. I think of Isaac Asimov's I, Robot and a hundred other science fiction novels. I know Kaikō to have been more akin to a cable-controlled model airplane, but still, I see her swimming there, alone, seven miles down. And it pains my heart to report her fate. On May 29, 2003, in heavy weather, the launcher came to the surface without the vehicle. Somehow, the vehicle had broken free.

This eventuality had been anticipated. The vehicle, without power, was designed to float to the surface and, once there, to send out a signal. But the car-size vehicle was never found. Her seven thrusters, her two onboard sonars, her seven video cameras, her high-definition still camera, and her two electrohydraulic manipulator arms with seven degrees of freedom and force feedback could no longer help her. All were lost. And her replacement was not designed for full ocean depth.

In recent years, some innovators have seen value not in maximizing depth, nor in maximizing capabilities, but in driving costs downward, in bringing ROVs within reach of budgets controlled by the likes of search and rescue divers, fish farmers, bridge inspectors, and even individuals such as writers interested in the deep sea. It is as if the innovators saw the aerial drone market explode and thought, *Here's an opportunity.*

I talk to Rusty Jehangir, founder of Blue Robotics, a company that provides components and even kits for do-it-yourselfers interested in ROVs. He started, he tells me, in aerospace, then moved to aerial drones, and then, for fun, tried to build a robotic boat in his garage using a surfboard. The parts he needed, especially the thrusters, were hard to find and "not simple or cheap to buy," as he puts it. While scrounging for parts, he realized that others were encountering the same challenges. He made what he calls a "three-phase brushless motor with the armature spinning in water." That is, he engineered a motor that did not have to be sealed. Water could come in contact with the armature. He could sell the motor at one-tenth the cost demanded by what he saw as the competition. And so in 2014, he launched an online funding campaign that presold six hundred of them, which led to an online store based out of his garage for do-it-yourself ROV aficionados.

The surfboard-based robotic boat was abandoned. The ROV business was taking all his attention, and in any case his robotic boat kept getting fouled in kelp.

He brought new components into his catalog one at a time to support cash flow, and by 2016 he had a more or less complete kit available. As the business grew, he rented more garage space, and by the time of our conversation in 2017, he had acquired seven thousand square feet of more conventional commercial real estate. By then he had enough business to keep seventeen full-time employees busy. If modified from the base kit, his robots could go to depths of five hundred feet.

Blue Robotics is based in coastal California, home to some of the best recreational diving in North America. I ask if he dives. "I don't," he says, "and now I don't have to."

While I disagree with his sentiment—after all, no kid wants to grow up to be a robot—I admire his spirit, his enthusiasm for his product line. And while I have no intention of giving up diving, of letting a robot have all my fun, I want an ROV of my own.

Uncharacteristically for me, this is more than an ordinary desire. It is an obsession. I cannot afford my own submersible or the boat to support it, but I might be able to buy a little robot capable of going way deeper than I can go with conventional diving gear. There will be no decompression requirement, no shivering, no expensive gases to buy, no risk of decompression sickness.

Jehangir tells me of plans to have a Blue Robotics booth at an upcoming diving equipment trade show in Orlando, Florida. And so I go, on the designated date, shopping for my own ROV.

Orlando's Orange County Convention Center overflows with diving equipment, ranging from wet suits to regulators, from fins to rebreathers. There are travel agents, dry bag manufacturers, and T-shirt makers. There are classroom lectures on regulator repair and scuba instruction. There is a pool for gear demonstrations.

Most of the attendees own or work for dive shops. They are here learning how to repair gear and deciding on which products to carry.

Scattered among the rows of vendors I find a dozen booths marketing or hoping to market ROVs sized for use from small boats. I talk to all of them, telling them, one after another, that I am a writer working on a book about humanity's presence in the deep sea, but also that I am interested in owning an ROV, in having one aboard the boat that my wife and I call home. Without exception, they are eager to talk about their products.

This statement, or something similar, from more than one vendor: "We were the first, and now everyone wants to copy us."

And this, from at least half of the vendors: "Ours is the most reliable."

And this, from a single vendor, clearly not talking about his company's product: "ROVs mostly suck."

But also this from another vendor: "The competition is brutal. Everyone is trying to show a profit. But really we would be better off seeing ourselves as part of a cooperative effort. Right now, we are all trying to build a market. We're trying to get recreational divers to buy ROVs."

And this message, repeated by three-quarters of the vendors: "We saw what happened with aerial drones, and we know there is a market."

Then a contrasting viewpoint: "These are not aerial drones.

Most people do not own boats, and let's face it, you need a boat to get much use from an ROV. And of those people who have boats, most do not use their boats in clear water. You need clear water to have any fun at all with an ROV."

From one vendor: "There is no reason to think that we will be limited by low visibility. We can put sonar on our product." He does not mention that the kind of sonar he envisions would cost far more than the ROV itself, and although it would sketch an image of whatever might be within range, that image would not be in full color or have the high resolution that consumers would demand.

From another, perhaps more grounded vendor: "If the water is not clear, if you can't see very far, then an ROV is not the right tool." Only I am not sure if he said "tool" or "toy."

Some of the ROVs are flat, somewhat wing-shaped, with horizontal thrusters to push them forward and vertical thrusters to push them up and down, designed to fly underwater in a manner similar to that of an airplane. Others are boxy, with as many as six thrusters, designed to move right or left, up or down, forward or backward, or simply to hover, to fly underwater in a manner similar to a helicopter.

Some, including their controllers and tethers, are as small as a large briefcase or a day pack, while others are the size of a small trunk.

Some are depth rated to a mere hundred or so feet, but most claim to work well to around three hundred feet.

Yellow is, not surprisingly, the most popular color, but white and blue are also represented.

All of them trail tethers, most of which are smaller in diameter than a household extension cord. Some have tethers that can be lengthened using supposedly watertight connectors. Most have tethers that go directly to a controller, but some have tethers leading to floating spools, and the floating spools are linked through a Wi-Fi signal to the controller. The operator can be some distance from the spool, and when the time comes, a signal can be sent telling the ROV to come back to the operator, towing its spool behind.

The controllers range from purpose-built units to smartphones to tablets to joysticks originally made for video gaming.

Some of the ROVs have cameras that can pan up and down, while others can pan only by moving the entire ROV. Some can carry lasers used to measure the size of, say, a fish or a piece of coral. Some can hold their depth and heading at the push of a button, while others rely exclusively on the operator's ability. Some have image stabilizers. Some can carry extra cameras.

All have at least one built-in light and work on internal batteries, and most can operate, or so claim the vendors, from two to four hours on a single charge.

The costs, for vendors trying to compete in the recreational market, range from about $1,800 to over $5,000. At least two vendors do not see themselves as competing in the recreational market, but rather in a market where buyers might need to inspect aquaculture infrastructure or bridge pilings or dams. Their machines, which seem far more substantial than the recreational machines on display, cost upwards of $30,000. In fact, it would be easy to spend $100,000 on a fully equipped ROV intended for, as one vendor described it, "light commercial work, such as inspections." And while still manageable enough to use from a small boat, they are bigger than the recreational ROVs, and one of the models is powered from the surface, giving it an indefinite run time underwater but requiring a tether almost as thick as a garden hose and a boat with a generator.

One of these vendors, a person who has been working with ROVs for decades—that is, from before some of the other vendors were even born—expresses surprise at the number of companies offering recreational ROVs. He worries that they might be in a self-destructive race for the bottom. Not the bottom of the sea, but the bottom of the bottom line. "No one," he insists, "can make a profit selling reliable ROVs for a few thousand dollars."

⌁

I fly a Blue Robotics ROV in a tank that is not much bigger than the robot itself. This ROV is built to hover, to swim like a helicopter flies. There is not much room to maneuver, but the thing is stable, responsive, easy to use. For my purposes, though, it is just a little too big. It would be hard to store in the limited locker space aboard our boat.

Later, I meet several vendors at the trade show's test pool, which is much bigger than the Blue Robotics display tank. It is about the size of a small backyard swimming pool.

The machines I fly in the test pool swim something like an airplane flies. Most vendors claim that it is possible to fly their ROVs right out of the box, no training required. If flying means bouncing them off the bottom and crashing them into the walls of the pool, they are right. But I blame the collisions at least partly on the operator: myself. I did not grow up gaming. I might need a little practice. And while the crashes harm my ego, they do not harm the ROVs or the pool.

Other newbie operators, ROV virgins like myself but young enough to have grown up gaming, fly figure eights and pop up and down like dolphins, totally under control.

⌁

In the end, I buy an ROV that was not represented at the trade show. It is a Deep Trekker, a model the manufacturer calls the DTG2 Smart. It is not, as a rule, marketed for recreational use, but for light commercial applications.

What drove my choice? The DTG2 Smart's small size, its

depth rating of 492 feet, the company's eight-year history making small ROVs, a removable manipulator arm, a high-definition low-light video camera that can not only pan up and down but can automatically maintain its angle relative to the bottom even when the ROV tilts, and the ability to maintain a depth and heading without relying on the limited skills of its operator. Its housing is aluminum instead of plastic. Its thrusters rely on sealed motors magnetically coupled to tiny propellers rather than the flooded motors used on most recreational ROVs. The magnetic coupling is important—when the propeller picks up a piece of fishing line or a strand of sargassum or a leaf of turtle grass, the magnetic coupling slips without breaking the propeller shaft or burning up the motor. And the DTG2 Smart has a unique system for going up and down, a patented internal pitch system. The outside of the ROV turns around an inner axis, pointing the thrusters upward to descend and downward to ascend.

On the downside, like most of the ROVs marketed for recreational use, it cannot truly hover. To stay in one place underwater, it relies on neutral buoyancy, supplemented by the operator's gentle but expert working of the controls.

The operator, in my case, has neither a gentle nor an expert touch.

On the controller, the left joystick moves the ROV forward and backward and turns it right and left. The right joystick controls the internal pitch system. Pushing the right joystick down while the ROV is moving forward sends her diving, while pulling it back brings her to the surface. Buttons on the controller can lock in her heading and depth, turn video cameras and lights on and off, change camera angles and thruster power levels, and open and close the manipulator's claw.

Practicing with my brand-new DTG2 Smart in the swimming pool at a mobile home park in Florida, I prove adept at repeatedly coming very close to colliding with walls. I bounce off the bottom more than once. I unintentionally bob along the surface.

An elderly woman—the only other person at the pool—watches silently, pretending to read. Eventually, she puts her book down.

"What is that thing?" she asks.

"An underwater robot," I tell her. "Good to depths of nearly five hundred feet."

"Oh," she says. "I thought you were cleaning the pool."

⌒

I practice in Florida bays and at shallow anchorages as we sail east and south in the Bahamas. I want to become at least competent before taking my DTG2 Smart into deep water, into a place too deep for me to conveniently dive.

With about ten hours of practice under my belt, I can fly over the seabed looking at sponges and soft corals and seagrass. I can quietly sneak up on a moray eel or watch a yellow stingray glide across the bottom. I can sit in the cockpit of our boat, warm and dry, while simultaneously swimming down our anchor chain to check for a firm hold on the seabed.

One night, anchored off a Bahamian village, I swim, once removed, in the darkness beneath our boat. I push the left joystick to the left to turn counterclockwise, starting what I hope to be a large circular search under the boat. My lights attract a cloud of plankton. The plankton attracts fish, each one or two inches long, shimmering, reflective, at one angle blue, at another angle red. Not one or two or three or even three hundred fish, but thousands of fish. The fish are so thick that I see nothing but their flashing sides and tails and heads. I turn off the lights and hover in darkness, giving the fish a minute or two to disperse. I turn the lights on. The fish recongregate within seconds. I push my joystick to the right, rotating the ROV clockwise, and the growing cloud of fish follows the ROV's light beam. I rotate counterclockwise, and the fish turn around. I am, I admit, delighted and ready for a deep dive.

⌐

The bathymetric chart for the area just west of Egg Island, which is itself just west of Eleuthera in the Bahamas, shows closely spaced soundings. Depths of 85 and 102 feet stand immediately next to depths of 1,378 and 2,460 feet, which are in turn next to depths of 6,118 and 7,349 feet. Our depth sounder shows not a smooth slope but a series of walls, a flooded staircase of cliffs. This sort of thing is not necessarily common underwater, but it is also not exceedingly rare.

In a light wind, we drop our sails over a spot on the chart between the 102- and 2,460-foot soundings. Our depth sounder blinks, meaning we are in more than 400 feet of water. We start our engine and steam shoreward until we pick up soundings. The depth goes from 380 feet to 120 feet over a horizontal distance less than the length of our boat.

Motor off. ROV over the side. The left joystick pushed fully forward for speed, the right joystick pushed fully forward for a vertical dive. My wife playing out tether as fast as she can.

The controller's video screen shows blue water streaked with sun.

But our boat, drifting, moves at over a knot, propelled by wind and current. The ROV operator—that would be me—does not have the skill to both dive and follow the boat. I surface and try again. Same result. The boat moves away too quickly to get depth. I do not even reach one hundred feet. I surface and try a third time.

I notice a small shark circling my ROV, but I cannot see it well enough to know what it is. A blacktip?

The third dive goes no better than the first and second dives. We retrieve the ROV.

On this day, I learn something that I already know: the importance of the surface vessel. It is not practical to anchor a sailing yacht in several hundred feet of water. It may be possible to hold our position with the boat's diesel engine, but sending the ROV down in close proximity to a spinning propeller is not an attractive proposition. I cannot simply hit a switch and stay put, held stationary by computer-controlled thrusters, as the mother ship of Kaikō could. Small sailing yachts do not have dynamic positioning systems. A deep dive will need an unusually still day or a different strategy.

That night, disappointed, anchored in a nearby cove, I study charts and nautical guidebooks and weather forecasts, looking for another opportunity.

Eleuthera's Ocean Hole, despite its name, sits a quarter of a mile inland, surrounded by the town of Rock Sound. It is a sinkhole reputed to be six hundred feet deep. More than one local tells me that it is bottomless. The water is neither blue enough nor clear enough to make it one of the famous Bahamian blue holes. It might be better called a green hole. Visibility at the surface is maybe fifteen feet. But an old man whom I meet on a Rock Sound dock tells me that divers going below two hundred feet deep have found crystal-clear blue water, somehow magically lit, full of sunshine. "Like da' brightest day, mon," he tells me.

Another man, gray-haired and stooped, says he has talked to the divers himself, and they have told him it is black as night down there. "'Bout like mud, mon," he says.

The tide rises and falls in the hole, suggesting a connection to the sea, but locals know of no one who has actually swum through, who has joined hole and sea. There are fish in the hole, snappers and angelfish and doctorfish. But the two old men agree that the fish were caught in the ocean and released here. Fish do not swim between the hole and the sea.

"The depth, mon!" one of them exclaims. "Too deep."

Small groups of tourists visit and loiter near the landing that has been built next to the hole. Most of them stare at the water for a few minutes, throw some bread to the fish swimming alongside the landing, and leave. A few swim close to shore. One or two venture out over the hole's center, or across its four-hundred-foot diameter, or even around its shoreline of limestone bluffs. Local kids, often in school uniforms, congregate here, too, hanging out, joking, sometimes swimming, but like most of the tourists reluctant to go too far from shore.

I lower my ROV into the water amid a mix of schoolchildren and tourists. I have the robot swim past the fish, shooting video, and then I head it toward the center of the hole and down. At forty feet, the ROV drops through a layer of suspended silt, a dense cloud that looks deceptively like the bottom but is not. Below the cloud, the water temperature readout on my controller increases from the high seventies to eighty degrees.

The camera points downward. Each time the ROV approaches the bottom—that is, each time my screen displays brown silt and decomposing leaves, which become visible at a distance of about three feet—I push the left joystick forward, leveling the ROV and moving it farther out toward the center of the hole.

At eighty feet, I—once removed, represented at depth by my DTG2 Smart, my ROV—pass through a second layer of suspended silt. Below it, the water temperature drops to seventy-four degrees.

I see mud. I push the joystick forward. I see more mud and push the joystick forward again. I inch along carefully, concerned about fouling my tether in the darkness. And as the ROV goes deeper and farther away from my station at the landing, I can feel the tether making it more difficult to steer. To turn now, the ROV has to fight the drag of several hundred feet of cable.

Imagine yourself in the water with a rope tied around your waist, trailing behind as you swim. The rope is, say, two hundred feet long. Each time you turn, the rope forms an arc, resisting your efforts. My little ROV, my DTG2 Smart, puts up with the same sort of resistance.

At a depth of 101 feet, the bottom slopes upward. Pushing the joystick forward takes me to shallower water, to more mud. If this hole is bottomless, I am in the wrong place.

I bring the ROV straight to the surface and back to the landing, and I try again, thinking that I may have missed the deep center of the hole. But again I reach just beyond one hundred feet.

Rumors of extreme depth may be exaggerated here. Or maybe the deep hole is more of a cave, difficult to find in the murk.

Taking the ROV from the water, I show it to some kids in school uniforms. Each in turn wants to pick it up, to feel its weight.

"How much dat costs?" asks one.

"A lot," I say, and then I let them see themselves on the ROV's video.

Reeling the tether onto its spool, I consider what I have learned. I am not disappointed, but I am becoming more realistic about what a cabled ROV can and cannot be expected to do. I am not alone in this realization. Anyone familiar with ROVs understands that tethers limit usefulness and maneuverability.

Bob Christ, CEO of a company called SeaTrepid that is in the business of underwater robotics, wrote the book on ROVs. This is not hyperbole. The name of his book, coauthored with Robert Wernli, is *The ROV Manual: A User Guide for Remotely Operated Vehicles.*

They discuss micro-ROVs and flying eyeballs and work-class ROVs and special-use ROVs, covering machines that weigh less than ten pounds and are armed with tiny thrusters and those weighing tons and carrying 250-horsepower thrusters that could, in principle, power a ski boat. They describe ROV business models and comment on the high failure rate of ROV companies. They expound on conflicts between manufacturers and users. They write of the characteristics of rigid polyurethane and syntactic foam, of compressive modulus considerations, of transverse stability, of propulsion systems, of magnetic couplings, of fluid-filled housings, of rotary joints, of Ohm's law and EMF effects, of rheostats and joysticks.

When I talk to Bob, he cannot hide his delight in all things ROVish. He is technical, yes, but he is also passionate. From his book: "Two critical groups of people have driven ROV history: (i) dedicated visionaries and (ii) exploiters of technology." Clearly, he falls squarely in category one.

We talk of many things ROVish. More than once he uses the phrase "design spiral." The parts of an ROV are so complex that a change in one results in the need for changes in others. Say, for example, that someone wants a longer tether. To pull the longer tether, that someone will need stronger thrusters. Stronger thrusters mean greater power requirements. Greater power requirements for ROVs powered from the surface often mean thicker tethers to decrease electrical resistance and increase insulation. All of this will change the buoyancy characteristics of the ROV, and thickening the tether will increase its resistance as it moves through the water. More flotation and still more powerful thrusters may have to be added. And the spiral continues.

Despite his passion, he is not optimistic regarding the potential for recreational ROVs to find a market. Flying deep with an ROV capable of pulling its tether effectively is tricky and expensive.

In the early days, he tells me, ROVs were controlled by switches and rheostats. A rheostat is a dial used to control power flow. If an ROV had six thrusters, it had six switches and rheostats. The operator—the pilot—had to turn one on and another off to move, say, to the right or left, and the operator was constantly spinning rheostat knobs, balancing power, trying to get the ROV to fly in something resembling a straight line. It was almost like playing a musical instrument. And the camera technology was expensive and just not that good. Available batteries did not have the power density—the electricity storage capacity per volume—to be used in ROVs.

Today, there are few switches and almost no rheostats. Instead there are joysticks. Inexpensive high-definition cameras that work well in low light have replaced their predecessors. Lithium-ion batteries provide energy density more than sufficient for at least the smaller units.

Comparing a modern ROV to, say, Dimitri Rebikoff's Poodle would be something like comparing a saturation diver to a diver in a suit and helmet made by the Deane brothers in the 1830s. Only more so.

"By the way," Bob says, "I knew Rebikoff." Rebikoff visited Bob's home and stayed there for a time, helping Bob and his father in their search for a lost German U-boat in the Gulf of Mexico.

Unrelated, he tells me of a committee he sat on in 2014. The members were tasked with considering the role of ROVs in the oil fields of 2025. "We were off in our predictions," he says. "I know now that what we envisioned for 2025 will be in place by 2020."

But also this: "Oh, God! This industry is ripe for progress. There is so much left to be done."

Which brings him to the topic of tethers. Tethers produce drag. Not a little, but a lot of drag. And the deeper an ROV goes, the more drag they create.

From *The ROV Manual*: "The highest factor affecting ROV

performance is tether drag." It is not the size or shape of the vehicle itself that is of most concern when it comes to drag, but that long leash headed back to the surface.

Bob both envisions and lives in a tetherless world. On behalf of a client—an investor who happens to manage a hedge fund but who sees vast sums of wealth to be made at the bottom of the sea—he just bought eight HUGIN autonomous underwater vehicles, or AUVs, the tetherless cousins of the ROV. The HUGINs can dive to twenty thousand feet on missions lasting sixty hours. While submerged, they are on their own, or can be, responding to preprogrammed commands along the lines of "Search this area using sonar by going back and forth forty feet above the seabed, but if you see something that looks like a treasure chest, stop to get pictures, and come back to the surface before your battery runs out so that we can recharge you, download your data, and send you back to the bottom."

"The HUGINs," he says, "have by far more survey miles behind them than any other AUVs on the market." But, he says, they are seldom entirely autonomous. A typical operator, justifiably concerned about the six-million-dollar machine just tossed over the side, follows the HUGIN from the surface, staying in touch through an acoustic signal. Compared to the information that can be carried through a tether, acoustic transmission provides limited control, but it can at least keep tabs on the robot and perhaps command it to surface if necessary.

Bob's vision for the HUGINs that his client has acquired diverges from standard protocol. They will go over the side one after the other, and they will be picked up, hopscotch fashion, one after the other, for recharging and downloading of data. The ship will stay in the general survey area, but on the bottom the AUVs will work on their own, following orders. The ship will not attempt to follow the individual machines as they go about their missions.

Right now, as I write, the HUGINs are out there, searching.

They are not, in this case, searching for a submarine or a treasure chest. They are in the Indian Ocean, eight orange robots, each about twenty feet long and three feet in diameter, swimming back and forth at something like four knots, scanning as much as 425 square miles every day, hoping to find Malaysia Airlines' elusive Flight MH370, missing since 2014.

From Oliver Plunkett, the CEO of Ocean Infinity, the organization in charge of the search: "Whilst there can be no guarantee of locating the aircraft, we believe our system of multiple autonomous vehicles working simultaneously is well-suited to the task at hand."

Indeed.

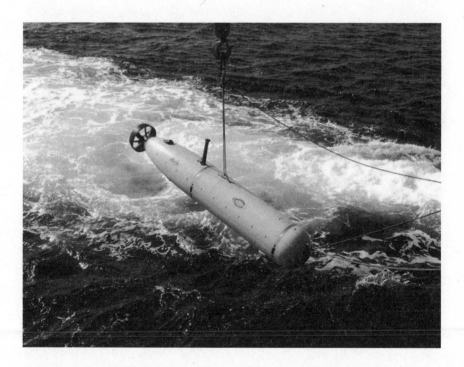

One of many kinds of autonomous underwater vehicles, or AUVs, programmed to operate underwater for long periods without direct control from the surface. This photograph was generously provided by Bluefin Robotics.

The HUGINs are not alone out there. Hundreds of AUVs have been deployed, including the REMUS, Autosub6000, ABE, Sentry, A9, A18, A27, and Alistar 3000. Most look something like torpedoes and are driven by propellers. But there are AUVs that look at least a little bit like living creatures. There is, for example, the RoboLobster, designed in part to crawl around in shallow water searching for mines. Another looks and swims like a jellyfish. A third imitates the appearance and movements of penguins.

And there are the so-called gliders, which coast forward and downward through the water until they reach a preprogrammed depth, at which time they adjust their buoyancy and float back up, reaching the shallows only to start a new glide path, cutting a sawtooth pattern up and down through the sea, collecting oceanographic data as they go. They neither need nor have propellers. The gliders, with their buoyancy chambers, are slow and cannot quite follow a straight line, but they can work for weeks or months on a single battery charge. A glider called the Scarlet Knight swam for seven months, crossing from New Jersey to Spain, in 2009.

Taking an entirely different approach to long-duration missions, Boeing has been building the fifty-one-foot-long *Echo Voyager,* a diesel-electric submarine without a crew. Like the submarines of World War II, like the nonnuclear submarines that continue to patrol many coastlines today, it runs a diesel generator on the surface to charge its batteries, and at depth it uses the batteries for propulsion. Lance Towers, speaking on behalf of Boeing, has called it, in discussions with the media, an "un-personed submarine."

And there is, or was, Nereus, the only ROV other than Kaikō with full ocean depth capability. Nereus, built by and operated by the Woods Hole Oceanographic Institution, went to the bottom of all bottoms—that is to say, the seabed of the Challenger Deep—on May 31, 2009. She was, to say the least, very different from most of her predecessors, and most would say far more advanced. She is, or was, before she was lost, a hybrid vehicle, part ROV and part AUV. Her twenty-five-mile-long tether was a thin strand of optical fiber, no thicker than fishing line, carried on spools aboard Nereus herself. Operators on the surface could control her through the tether, but she could also be cut free, sent off on her own.

But, ultimately, she was lost at sea, failing to return to the surface in one piece after a dive to 32,500 crushing feet, shy of her deepest depth but by no means shallow. When she was lost, on May 10, 2014, Ken Kostel of the Woods Hole Oceanographic Institution posted a simple but telling message. "We lost Nereus today," he wrote. "It was always a possibility, given where it was designed to go. The fact that it was lost doing what it was intended to do in no way lessens the sting. Nereus is gone." And there was this, from the institution's director of research, Larry Madin: "Extreme exploration of this kind is never without risk, and the unfortunate loss of Nereus only underscores the difficulty of working at such immense depths and pressures. Fortunately there was no human injury as a consequence of this loss. WHOI scientists and engineers will continue to design, construct and operate even more advanced vehicles to explore and understand the most remote and extreme depths of our global ocean." And there were notes of condolence, many suggesting that their authors had come to think of Nereus as something more than a robot. At least one was signed "with love." And one came from James Cameron, who started his note with the words "I feel like I've lost a friend."

While humanity does not currently possess a robot capable of reaching the deepest of ocean depths, technological advances continue. Just leaving the starting gates are AUVs that can download data, upload new mission orders, and recharge their power systems without surfacing. Bob Christ calls them "sleep-on-the-bottom AUVs." The dream is to have charging and data-exchange systems that do not require the AUV to be physically connected to anything on the surface. Instead, the AUV would swim to a submerged control station. There it would transmit data and receive new orders, and it would be charged through what amounts to induction. The sleep-on-the-bottom AUV would have to surface only occasionally for servicing and perhaps to have the barnacles scraped off its hull.

Sleep-on-the-bottom AUVs will change oil fields. Not one but many might support the wells and pipelines of an offshore production facility. Engineers ashore would receive the AUVs' data and occasionally send out new orders.

Meanwhile, I will continue to practice with my DTG2 Smart. I will hope that those who do not believe in a recreational market for ROVs will be proved wrong. I will do what I can to overcome tether drag, to avoid the problems of currents and wind and drifting boats. And, eventually, I will dive my little robot to depths well beyond those that I can conveniently reach with diving gear.

Chapter 7

AN OCEAN IN NEED

More than a year has passed since I first met my longtime hero Don Walsh. It has been a period of revitalizing my love for the underwater world through free diving, of reminding myself of what it took for divers to reach extreme depths while breathing what George Bond called synthetic atmospheres, of learning about submarines and robots, of acquiring a robot of my own. And now, near the end of this year, I find myself in Clarence Town Harbor, on Long Island in the Bahamas.

Along with my wife aboard our old sailboat, I wait out a spell of bad weather, hiding from high seas. Waves crash over the surrounding reef before running out through the mouth of the harbor, setting up a three-knot current, establishing a constant surge. At night, when we should be sleeping, we listen to the sounds of the boat straining at its anchor in water moving as fast as that of many rivers, and it feels as if we are under way, moving forward through a heavy chop. It is not a feeling conducive to sound sleep.

But Clarence Town Harbor is not without its charms. Foremost among these is Dean's Blue Hole, just a few miles down the road, a flooded sinkhole. Unlike most, it is reasonably deep, easily accessible, more or less free of currents, and usually clear. Its bottom sits 663 feet below the surface.

If there is a place where it should be easy to take my little ROV to its maximum design depth, to almost five hundred feet, this is it. And among free divers it is renowned as both an ideal location for training and the site of the Vertical Blue international free diving competition. William Trubridge set a world record here by swimming on a single breath, without fins, to a depth of 335 feet. And, as it turns out, Trubridge lives just around the corner from Dean's Blue Hole. He runs a free diving school here.

⌣

During the past year, my wife and I have tried to keep up our free diving skills, but just as it can be challenging to find the ideal location to easily fly a small ROV to reasonable depths, finding a good spot to train is not as easy as one might hope. More often than not, we have been diving on our anchor in ten or twenty feet of water or on reefs no deeper than sixty feet. Neither of us has been past eighty feet on a single breath since leaving Honduras many months ago.

We go to look at Dean's Blue Hole with plans to return the following day fully equipped. Limestone cliffs and a sandy beach surround three sides of its deep water, while the fourth side is open to a lagoon protected by a reef. We have our masks with us, and so we decide to wade into the hole. Two steps off the beach put us in ten feet of water. A couple of quick strokes take us over a patch of dark blue that screams depth. A few more strokes place us next to the platform secured in the middle of the hole, floating sixty stories above the bottom.

And so we make a few quick dives to quench that strange free diver's hunger for depth. On my third dive, I reach 101 feet without even trying. My wife, who does not have a depth gauge with her, goes somewhat deeper, reaching a disk that would typically hold tags used during competitions.

Our training has stuck with us. In our fifties, far from athletic, we have reached depths that neither of us thought possible a bit more than a year ago. And we have done it without pushing ourselves.

It is late in the day, so we stop after just three dives each, looking forward to returning the following day fully equipped with wet suits, weights, fins, and, for afterward, the ROV. I have plans to beat William Trubridge's record with my DTG2 Smart.

Free diving champion William Trubridge wrote *Oxygen: A Memoir* while here in the Bahamas. It includes some memorable lines. My favorite: "When we can respond to suffocation with equanimity, we are truly on the path towards maximising our breath-holding and hence our aquatic potential."

He wrote, too, of Dean's Blue Hole, describing it as "a brooding liquid mystery that swallowed the light from the sky," as a place "that more than anywhere else on the planet would enable me to unlock the aquatic potential that was in my species."

He described his early dives: "The perpetual twilight of the depths beyond the rim of the Blue Hole became familiar and soothing." Before the end of his first two weeks on Long Island, he was soothed by depths greater than 230 feet. In a bit more than a decade, he would break his own world record there.

But he also wrote of trade winds blowing "unrelentingly," sometimes churning up sand, destroying the visibility, driving rafts of sargassum into Dean's Blue Hole, the floating gulf weed stacking up until it was thicker than he was tall. But this, it seems, hardly ever happens. And so I was confident that we would be able to dive Dean's Blue Hole, my wife and I and my surrogate, my stand-in, my DTG2 Smart.

But today, the day after our arrival, driving down to Dean's Blue

Hole, we have to cross a large pool of water in the sandy access road. The pool is full of sargassum. My confidence dwindles.

Turning the final corner, looking across Dean's Blue Hole, we see fifteen-foot-tall ocean waves crashing over the reef. What had been a calm lagoon is now a washing machine. And we see a spiraling mass of sargassum centered over the hole itself and piled up on top of the free diving platform. Our excitement, our anticipation, drains away. Walking out into the sargassum, pushing it aside, we see that what had been clear water just yesterday is dark now. Looking down through the opening that we make in the rafted weeds, we cannot see our own feet.

There will be no diving today. Our fins and weights and wet suits will remain dry. My little DTG2 Smart will stay nested in its heavy plastic storage box.

I call William Trubridge. I tell him what is going on at his blue hole. He is surprised. He says that it normally takes a hurricane to throw sargassum up past the parking area. And he agrees to an interview.

~~

Back aboard our boat, taking our lumps from the surge and current in Clarence Town Harbor, I read a press release. The headline: "Russia's Nuclear Underwater Drone Is Real." While I have been busy writing and free diving, while I have been interviewing, jotting down notes, and thinking about humanity's presence underwater, the Russians have been testing a new kind of AUV, an autonomous underwater vehicle powered by an onboard nuclear reactor.

The thing is called Ocean Multipurpose System Status-6, or, by the CIA, Kanyon. It is, according to reports, about eighty feet long and thirteen lucky feet in diameter. It can, as far as media reports are concerned, travel unsupervised for 6,200 miles at speeds above sixty miles per hour and depths down to 3,280 feet. It carries or may carry a one-hundred-megaton device, a bomb something like six thousand times more powerful than the bomb dropped on Hiroshima on August 6, 1945. But Kanyon's bomb is not meant to be dropped on a city. It is designed to be set off underwater, near some enemy's coastline. Its explosion would, according to some sources, create a tidal wave, a tsunami, that could reach sixteen hundred feet high. And, just to make things a little more interesting, that huge rolling, crushing, unstoppable wall of water would be full of radioactive isotopes.

I can only hope that the Russians have far less luck with their AUV than I have had with my ROV.

William Trubridge—he goes by Will or William, he tells me— lives in one of the three homes on the side of a sand road a few miles from Dean's Blue Hole. There is a training pool behind the house. It was paid for, he jokes, with money from Steinlager, the beer company that, for a time, sponsored his dives.

Dramatic footage in a Steinlager video advertisement shows him swimming into Dean's Blue Hole, wearing only a bathing suit and goggles, going downward with a Tarzan stroke. Shadowy voices, some with Bahamian accents and some with the inflections of his home country, New Zealand, fade in and out of the soundtrack.

"This could be your last breath," says one.

"No, we don't go down there," says another, and, "That hole is dangerous."

A woman's voice proclaims, "You might never resurface."

A male voice: "Your body won't cope," and, "Humans are just not meant to go that deep."

In the context of the mental discipline required to pull off extreme dives, William Trubridge sometimes talks about "the chattering monkey inside my head" and thoughts that pop up like "the voices of attention-seeking toddlers," all distracting him, all doing their level best to keep him from doing his level best. A big part of deep free diving involves ignoring thoughts of failure, imaginary voices suggesting that the surface might not be a reasonable place to leave so far behind, that a breath of air sure would lighten the mood. The voices in the video advertisement sound like those of the chattering monkey, of the attention-seeking toddlers.

But, of course, Will surfaces safely. It would not be much of an advertisement if he had not. So despite the chattering, it was not his last breath. And then the words "Born to defy" drift across the screen just before a bottle of Steinlager beer appears, followed by "Keep it pure."

The advertisement affects me in two ways. It makes me want to drink a Steinlager, and it makes me want to dive Dean's Blue Hole.

Inside Will's house, a guitar leans against a wall in the lightly furnished, well-lit living area. He offers a choice of organic juices or water. Steinlager beer is not, for the moment, on the menu.

A copy of his recently released *Oxygen: A Memoir* sits on a table. He tells me that rights have been sold for translations into Spanish and Chinese.

"Free diving," he says, "is blowing up in China."

He is tall, lean, and dark-haired, and he wears shorts and a green T-shirt partly covered by porpoises. He comes across as intense but not tense, as possibly just a little bit uncomfortable with interviews, as someone who leads a very rich inner life.

"When you were born," I say, "the free diving record was 331 feet. It was done by someone riding weights down and riding a buoyed sled up. Since then, you have been deeper than that, swimming with no fins. So what's next? What's going to happen in the coming thirty years in the world of free diving?"

He emphasizes that sled diving—descending on a weighted sled and ascending with the assistance of a gas-filled balloon—is no longer an accepted part of the sport. It has claimed too many victims. He considers swimming without fins to be the purest form of diving. It is what matters most to him and, he thinks, to many others.

He does not foresee his records being smashed. Broken, yes, but smashed, no. Someone—maybe he himself—will add a few feet. But he does not believe that humans, unassisted, will go much deeper.

His words remind me of the Italian doctors who told sponge diver Stathis Hatzis that he should not attempt to retrieve the lost anchor in 1913, and those who told Raimondo Bucher that his 1949 dive would be impossible, and the voice on the beer commercial saying, "Humans are just not meant to go that deep." But at the same time, I suspect that he is right. And, of course, I hope that he is wrong.

Without prompting, he mentions the possibility of nanotechnology augmenting human hemoglobin. Nanoparticles, properly designed and injected into the bloodstream, could help store and transport oxygen. Research in this area, he says, has already started. If it works, depths will be extended. It would not be the same as what he has done with unaugmented blood, but it might appear the same. An apparently unassisted diver could swim, without obvious technological aids such as weighted sleds and balloons, to, say, five hundred feet. Or deeper.

With all the research I have done, the people I have talked to, the papers I have read, I have not heard of this technology. I make a note to follow up.

The public, he tells me, is not very interested in new records for scuba diving or even saturation or submersible diving. And certainly, the public is not interested in robots. People want purity. They want to see a person swimming unfettered by technology. They want to see a fellow human doing something extraordinary without assistance.

I ask if he has any idea how many times he has lost consciousness or lost muscle control during training and competitions. He does not. But he has had, until recently, a six-year stretch during which he did not lose consciousness underwater. And that stretch was broken, he contends, because he was fighting some sort of virus. His mind did not account for the fact that his body was not performing at one hundred percent that day.

It is not only his words that strike me but the way he says them. For a person describing multiple personal experiences with loss of consciousness and loss of muscle control, apparently too many to count and at least some of them at depth, he seems very casual. It appears that he can respond to suffocation with equanimity.

But at times he has feared for his life underwater, even if those fears were not, he believes, realistic. He thinks they were triggered by the combination of nitrogen narcosis and an accumulation of carbon dioxide.

He recently suffered from decompression sickness, rare but not unheard-of in free divers. He had been diving repeatedly to depths around 120 feet, up and down, up and down. And at the end of the day, his right side felt numb. A bubble or bubbles in his body had done something to his nervous system. He flew to Nassau and underwent four days of treatment in a decompression chamber.

That experience suggests to him that longer surface intervals and slower ascents might be in order during days of repeated dives to depths beyond one hundred feet or so. But, in fact, no one really knows. Neither John Scott Haldane nor his successors considered free divers.

Which prompts me to ask if he has ever used scuba. He says he has not.

I ask if he knows of any practical uses for free diving. He mentions, of course, the commercial lobster harvest, which is done in parts of the Bahamas and elsewhere by free divers. He says that some abalone and sea urchin divers free dive by choice in New Zealand, as a matter of efficiency. Ama divers in Japan, although dwindling in numbers, still ply their trade, mostly harvesting seafood. But really his interest seems to lie with free diving as a means to explore human capacity, to test and expand notional limits, to do what seems to be undoable, and to do it—although he does not put it this way—gracefully and graciously.

I ask if he knows that Lucayo Indians—the native people who once inhabited the Bahamas and swam in Dean's Blue Hole—dived for pearls and were sometimes enslaved by the Spanish, who had their own special hunger for pearls. Later, after the Lucayo had been killed off by malnutrition, overwork, disease, and murder in various forms, African slaves were forced to dive. Some of their descendants may be Will's neighbors today.

"Those only are accounted complete divers," read one writer's words from 1750, "who kept themselves under water till the blood gushes from their eyes, mouth and nose." Another writer's words, from 1777: "Those who have disappointed their master in their catch of pearls, or who are contrary, they keep in these dormitories or prisons, grills, and cells, and they punish them by beating and flogging them in a cruel and savage manner."

This seems to be news to Will. I promise to send him a reference.

And then, at my request, Will talks quietly about his future. Many people ask what he intends to do next, maybe implying that he is aging out, that he cannot break records forever, that he cannot build his entire life around running his free diving school on Long Island.

Repeatedly breaking world records, even if it can be done into middle age or even old age, does not guarantee one's financial fortunes. He has other irons in the fire, irons more likely to support a family. Some he does not want to mention, but he does bring up the possibility of a career in motivational speaking, in coaching influential people in the corporate world, perhaps training them through free diving to ignore the chattering monkey of doubt and to quiet the voices of attention-seeking toddlers.

I cannot help but envision and hope for company presidents and directors secretly practicing breath-holds during board meetings.

But Will would also like to become more active in conservation. His issues, at least for now, are plastics in the ocean and the unrelated ongoing loss of New Zealand's Māui dolphins. "There are less than a hundred left in the wild," he says, "and they keep dying in fishing nets." His intonations, the look in his eyes that accompany these words, indicate his depth of feeling, his disgust with the situation.

Toward the end of our conversation, he says that he hopes I will convey the accessibility of free diving. "It is something that everyone can do," he says.

Yes, I think. *Even I can do it.* And I have seen people in their seventies take up the sport, reaching thirty or forty feet after a few days of instruction.

When I ask him if I can use certain quotes from his book, he says that he would be happy to see that happen.

From his book, a few lines of verse sum up not only William Trubridge but thousands of others:

I have a relationship with the depths
They beckon me beyond my means
Cold, dark, vacant pressure
Forever night, muted dreams.

Following up on Will's mention of nanoparticles that could mimic hemoglobin, which could enhance the abilities of free divers, I talk again to Dr. John Clarke, the scientific director at the U.S. Navy Experimental Diving Unit in Panama City, Florida.

"Right now," he tells me, "it's more in the range of science fiction, but a lot of thought and money will go into making it science fact, if possible."

And then he tells me that he was recently asked to prepare a briefing on the potential for developing artificial gills for humans, a topic that has attracted interest, resulted in a few experiments, and been dismissed as impractical several times over the past few decades. Likewise, there is interest in technologies that would allow humans to breathe the very thin, almost oxygen-deprived atmosphere on Mars.

"So apparently," he says, "science fiction is becoming more mainstream all the time."

Which leaves me wondering just how far along the research on nanoparticle hemoglobin supplements might be, and if John's response—his "it's more in the range of science fiction"—might not have been a clever way of saying, "Sure, but it's classified."

All of my thinking about free diving, exotic gas mixes, submersibles, and robots has left me wondering about conventional scuba diving. Despite current popularity, will it soon become just a little old-fashioned, a bit passé? Some in the recreational diving community push consumers to look beyond ordinary scuba diving. At least one training organization guides its students to refuse to breathe air underwater, to always dive with synthetic mixtures of nitrogen and oxygen or, in slightly deeper water, synthetic mixtures of nitrogen, oxygen, and helium, all with the intent of increasing bottom times and decreasing the risk of decompression sickness. Free diving organizations do not see their sport as an extension of scuba diving, or vice versa. And there are those who have told me that there is no reason to scuba dive at all, that their robots can dive for them.

So I talk to Alex Brylske, currently a professor of marine science and technology at Florida Keys Community College in Key West and possibly the most published writer in the world of scuba diving. He has, in his career, trained thousands of scuba divers.

He tells me that no matter what else changes, conventional scuba diving is here to stay. It is too easy, safe, and convenient to be replaced by rebreathers, by ROVs, by nanoparticles that mimic hemoglobin, by artificial gills.

And, of course, he is right.

But there is also this, from my trainer and coach, Mark Rogers, who taught thousands of young people to scuba dive but who now teaches nothing but free diving. "In the last decade," he says, "free diving has snowballed in popularity, particularly in areas where young people spend time in the water. Scuba diving is still one of the best ways to experience the sea, but let's face it, it's a leisure activity. Free diving is a mind and body sport, and its nature is simplicity. Similar to other popular activities like yoga and rock climbing, it rewards calm awareness, discipline, and judgment. But it also involves superhero costumes, specialized equipment, exotic locations, impressive depth numbers, and cameras—cameras everywhere. The sport has grown dramatically, but it's probably just getting started. I don't know if 'Instagrammability' is a word yet, but free diving sure has it."

And from Herbert Nitsch—record-holding free diver in multiple disciplines, the man badly injured while riding a weighted sled to 831 feet on a breath of air—with regard to the future of diving, with regard to what comes next: "Each time I think

I've reached a limit...there is a door...it opens...and the limit is gone."

~

During this past year of thinking about and writing about people in oceans deep, about humanity's presence beneath the waves, I have drawn from my own experiences, both new and old, but more importantly from the experiences of hundreds of others. I have more than seventy thousand words of scribbled notes from interviews and more than five hundred references. I have covered, in broad strokes, the story, as I see it. And while I could go on forever with interviews, with references, with tales of innovation and bravery, with reports of newly released technologies, it is time, in the course of the writing trajectory, to move this book toward its end.

But I know there is an important missing piece. There is James Cameron, the filmmaker who was the third and most recent human—at the time of this writing—to visit the bottom of the Challenger Deep. He is important in his own right, for his accomplishments underwater, but he is also important because my hope as a writer was to bring this book full circle, to start and end in the Challenger Deep.

Most people—that would be those of us who do not live in hermitages or isolated caves—have come across James Cameron's films. *The Terminator, Aliens, The Abyss, Titanic,* and *Avatar* are among them. But more relevant to my purposes, Cameron has taken some of the earnings from his films and used them to explore and promote the underwater world. By the time of the Gulf of Mexico oil spill in 2010, he was enough of an expert to feel comfortable publicly criticizing BP's response efforts, saying, "Those morons don't know what they're doing."

Two years later, Cameron took the *Deepsea Challenger* to the bottom of the Challenger Deep. Where *Trieste* had used a balloon full of gasoline for buoyancy, the *Deepsea Challenger* used syntactic foam, an essentially incompressible buoyant material that did not exist in 1960. And the pressure capsule of the *Deepsea Challenger*—the place where the pilot sits—was smaller than that of *Trieste*. It was a sphere with a diameter of forty-three inches, meaning room for one. Cameron dived alone, hunched in his tiny sphere, controlling the submarine's thrusters, instruments, and sampling devices in a space too small for outstretched arms. The dive, including descent, ascent, and time on the bottom, lasted about eight hours, a full working day. James Cameron and his submersible *Deepsea Challenger* reached a depth of 35,787 feet, just ten feet shy of the record depth found by *Trieste* more than five decades earlier.

Don Walsh was aboard the support ship. After Cameron surfaced, Don asked if he had come across any of *Trieste*'s iron shot, its ejected ballast, jokingly reminding Cameron that he—Don—had been there first. Later, Don talked to a *National Geographic News* reporter by telephone. "That was a grand moment," Don told the reporter, "to welcome him to the club."

It was a club of three—Don, the late Jacques Piccard, and now Cameron.

Don later recalled talking to Piccard immediately after their 1960 dive and agreeing that it would probably be between two and five years before anyone went back. It turned out to be fifty-two years.

While I was not able to reach Cameron, my fallback plan involved reaching someone involved with China Shipbuilding Industry Corporation's efforts to build a submersible capable of reaching that deepest of all depths, capable of qualifying crews to join the most exclusive of all clubs. But the Chinese, like Cameron's people, did not return my messages.

Which, it turns out, may have been fortunate. It changed the structure of this book, moving it away from something that circles back on itself to a narrative with an unexpected ending, an ending that surprised even me. It comes from an interview with another hero of mine, a person whom I have always seen as sitting next to Don Walsh in my personal pantheon. That person is Dr. Sylvia Earle, who set a depth record for women in 1979 by taking an atmospheric diving suit to 1,250 feet.

~~

Before moving on to Sylvia Earle and how she changed this book, I should mention that James Cameron and the China Shipbuilding Industry Corporation are far from the only interviews I failed to obtain. There is, for example, underwater archaeologist Robert Ballard. There are two manufacturers of autonomous underwater vehicles, both of which expressed initial interest but then stopped responding. There is the Woods Hole Oceanographic Institution, which would not grant me an interview but instead helpfully referred me to their admittedly well-done website.

I should also mention those whom I interviewed, including old friends, whose words did not find an obvious place within my words, within my book.

There are, for example, recollections from my old friend Dick Mason, a onetime British diver and welder. I last saw him in person when I was twenty-five years old. We were aboard a diving support vessel in the South China Sea, off the coast of Borneo. He was older than most of the divers onboard, and he was respected not so much because of his age but because of his welding skills and his personality, which included a quiet sense of humor.

The dives we were making required several decompression stops in the water, an ascent to the surface, and immediate repressurization in a chamber to finish off-gassing, using a technique called surface decompression using oxygen, or $SurDO_2$. As often happened, although it should not have, Dick's dive did not go as planned. Midway through the ascent from his last in-water decompression stop, he was delayed. This was through no fault of his own. The dive stage—his elevator to the surface—was powered, or supposed to be powered, by compressed air coming from the ship, but the ship's compressor had been turned off for maintenance. Although at this point Dick did not feel any symptoms of bends, his decompression, in the jargon of the trade, had been compromised. Nevertheless, he surfaced and was repressurized as if nothing had gone wrong. He completed his decompression in the chamber as if nothing were amiss. He stepped out of the chamber. The rest of the crew, including me, left him alone to shower and put away his gear while we headed to the ship's galley for a meal.

By email, Dick shares with me his recollection of those events so many years ago. He remembers, he writes, "a distinct sensation of a cloud passing over the sun, or was it the brain?" His logbook, filled out later, notes, "Stomach cramp, to an odd lower lumbar pain, disturbed vision then numbness of legs."

Somehow he made it to the ship's galley. As I recall that day, he sat down but was nonresponsive, insensible, almost comatose. The cloud passing over his brain prevented him from asking for help. One of us—Dick says that it was me—noticed that something was wrong. I do recall snapping my fingers in front of his face and noting his total failure to blink. And then two of us half carried him—at this point he could not walk on his own—back to the chamber.

We treated him using, again in the jargon of the trade, Table 6, a standard treatment schedule, a combination of times and depths designed for severe decompression sickness. His symptoms disappeared. He was sent ashore to be examined by a physician. He was declared fit to continue diving.

Not long after, he made a very shallow dive, working in six or seven feet of water with no ill effects. And soon after that, he was at forty-five feet. When he surfaced from this second dive, his symptoms surfaced with him. He was treated again, but it was the end of his career.

Knowing that he had a young wife and child to support, he became a teacher, taking a position with a school in Malaysia, where he stayed year after year, educating young men in the niceties of literature and grammar. He spent untold hours not decompressing but grading papers.

I ask him, more than three decades after the fact, if he regrets working as a diver, if he harbors any resentment over what was an avoidable accident, its cause firmly rooted in the carelessness of others.

"No," he replies. "Is it not like a soldier accepting the risk of death or being maimed in pursuit of the calling? It may be said we are all sons of Martha, accepting the drawbacks of our work." I have to look up the allusion to Rudyard Kipling's poem "The Sons of Martha," an ode to all those who work in the background, keeping things running without glory, without recognition.

And then he adds: "Regret? No. But to this day inconsolable sadness."

As a second example, I have a few words offered by another old friend, an American diver who still earns his living underwater, despite his advancing years. He was already an experienced working diver when we met, when I was eighteen years old. He is one of the few commercial divers I worked with who stuck with it, who did not move on to a more rational career. Knowing him to be a person of fearlessly strong opinions, I ask him if safety has improved for commercial divers. His reply, in essence: "More paperwork, safety plans that cater to the lowest common denominator, and young divers who don't know anything about construction." He may be right. Accident statistics for working divers have dramatically improved over the past several decades, but this may reflect the increased use of robots and the decreased reliance on saturation, and even more so the reluctance to allow short, deep dives with marginal decompressions.

And there is a third example, thoughts provided by another old friend, Mark Gittleman, who, with a strong nudge from his wife, gave up commercial diving when there was still time to be trained as an engineer. After trading his diving helmet for a marketable degree, he spent years working for the space division of a company that had, like him, started underwater. The idea was that there was money to be made in the common ground of harsh environments.

"The economic cycling of the space and subsea industries from the early 2000s through today," he says, "has meant that technical talent has moved in great numbers between the two industries."

And this: "Houston is now home to several robotics companies that are focused on applying NASA and other non–oil field technologies and techniques to deepwater oil field operations."

"I believe," he says, "that the traditional ROV is nearing the end of its reign as the king of deepwater subsea operations. Hungry, technically sophisticated new entrants are pulling technology from the space, automotive, and computer industries, and they are working on significantly cheaper and more capable systems with novel operational approaches." He is thinking of sensors, batteries, power management systems, position control algorithms, and even manufacturing processes.

"The traditional ROV will be replaced with autonomous and semiautonomous vehicles with innovative new power sources and batteries," he says. And, like Bob Christ, he talks of robots that will be controlled from shore and that will function without the need for expensive support vessels. "Remote commanding from shore," he says, and then adds, "Think Mars rovers," and "Think driverless cars with remote commanding capabilities." He envisions what he calls "resident underwater vehicles"—that is, the same kind of underwater robots that Christ told me about, the kind that will not return to the surface to recharge their batteries or to download data or to take on new orders.

He talks also of tiny satellites—"micro- and even pico-satellites"—using "swarm technology." The vision for space is one of hundreds or thousands of satellites communicating with one another and working together toward a common goal. And in Mark's view—and Mark is a practical-minded person who does not waste time dreaming of the impossible—micro- and pico-robots and swarm technology may soon be coming to a sea near you.

For some time, months in fact, I have been trying to interview Dr. Sylvia Earle. Her accomplishments go far beyond her record-setting dive in an atmospheric diving suit. Sometimes called Her Deepness, she led a group of female saturation divers in a 1970 dive in the Tektite II habitat, one of Sealab's successors, staying underwater for ten days. She was the first woman to hold the position of chief scientist at the National Oceanic and Atmospheric Administration. She designed and continues to design, build, and consult on submersibles and underwater robots. Since 1998, she has been a National Geographic Explorer-in-Residence. She started the exploration and conservation organization Mission Blue with prize money from a TED Talk. And she has worked with, among others, Don Walsh.

Our paths have crossed several times. But Sylvia's life, or so it seems to me, has become one of intentionally crossing paths with as many ocean lovers as possible. Her life has become a mission of ocean promotion. Most recently, with a group of journalists, I listened to her describe Mission Blue and what she calls Hope Spots.

"Mission Blue," she said, "is about understanding the oceans." It is also about inspiring people to protect the oceans. Hope Spots are places in the world that she considers especially important to the health of the oceans. Some are formally protected and others are not. Some are easy to reach and others are not. Some are known for excellent diving and others are not. Some are deep and some are shallow. Currently, more than ninety of them are scattered around the planet, including the Andaman Islands, the Bay of Fundy, the Gulf of Guinea, the Sargasso Sea, the Tasman Sea, Spitsbergen, and Lord Howe Rise.

Getting people underwater, she told us, may be the best way to inspire them to care about the oceans. Take people to Hope Spots, get them underwater, get them excited about protecting what is left of the oceans.

But there was also this: "While we have gained access to the oceans, we are losing the oceans."

After her presentation, we talked very briefly, one-on-one, and she took my notebook from me to write down her email addresses and phone numbers, encouraging me to contact her for a more in-depth interview. But my attempts to reach her went unanswered. At first I thought she must be too busy to respond, but as time went on, what I first viewed as my persistence morphed into something more akin to pestering, to stalking. I began to think that I might have said or done something to irritate her. Or maybe she looked into my background and discovered that I had worked as a commercial diver in oil fields and continued to work closely with the oil industry even after I became a biologist. Or maybe, being well into her eighties, she has more important tasks at hand than talking to a writer. Little things, such as saving the oceans.

Had she been almost anyone else, I would have given up. But Sylvia Earle is, well, Sylvia Earle.

As my sailing life continued in its parallel path to my writing life, I tried to reach her from Punta Gorda and Key West and

Marathon in Florida, and then from Marsh Harbour and Elbow Cay and Eleuthera and Clarence Town and Great Inagua in the Bahamas, all without success.

And now I decide to try one last time. I am in Haiti, anchored in front of a small village on the southwest coast. I promise myself it will be the last time, that I will leave her in peace after this attempt.

In what may have been a rare unguarded moment—she may have been expecting a call from someone else—she answers. She says she remembers me. She agrees to a short impromptu interview.

I ask her about the future of diving.

"The future I would like to see," she says, "is one with more divers, with more voices encouraging others to see parts of the world they will otherwise miss."

Divers, she suggests, are ambassadors. "Divers have the opportunity and the responsibility to share with others."

And this, echoing William Trubridge: "There is a place underwater for everyone."

But it is more than just scuba diving that she envisions. She sees a future in which submersibles will, in her words, "democratize access to the deep sea." Her organization is building submersibles intended for general public use, good to depths beyond three thousand feet. These are not submersibles that will sit on a megayacht or be used only by scientists. They will fit inside standard shipping containers, ready to go anywhere in the world that is serviced by containerships, ready to be loaded onto standard trucks, and from there to be moved to any vessel of opportunity or even to a dock close to deep water, where, like Karl Stanley, she could dispense with the need for a surface vessel altogether. A dock on Palau comes to mind.

Her intended passengers: schoolchildren, fishermen, corporate executives, civic leaders, politicians, everyone.

"Surprise, surprise," she says. "Most life on earth is living in the dark under pressure. You can't care if you don't know."

While working on this book, I have been struck by a very slanted gender ratio. Most people working underwater are, it turns out, male. Not all of them, to be sure, but most. So I urge Sylvia to talk about herself as a role model for other women.

She does not seem to be interested in this line of conversation.

"You are," I say, trying to get at the question from another angle, "a very accomplished diver and probably the world's most accomplished female diver."

She stays on point. "I don't think of myself as 'most' of anything," she says, "except maybe 'most frustrated.' I am frustrated by apathy about nature, about the sea."

She was around before plastic debris buried beaches and formed drifting islands far out at sea. And even before the oceans were, in her words, "clear-cut" by the fishing industry.

"When you take action, you get results," she says. "The wrong actions give the wrong results, but the right actions give the right results."

I mention that I am in Haiti, and that here the fishermen have no choice but to catch whatever they find, to take everything. In their absolute poverty, they have no other option, or, if they do, it will involve missed meals in an already sparse diet. She rejects this notion wholeheartedly.

"Of course they have options," she says. "They know better. They are like farmers eating their seed stock." They should, she says, find another way. They need to use their brains, to be creative, to think about how they can use the ocean without destroying it. Poverty is no excuse for stupidity.

My interview with Sylvia Earle is not exactly what I had envisioned. I had wanted to talk about diving technology, but Sylvia steered the conversation to conservation. This book, when I set out, was not meant to be about conservation. It was supposed to be about people underwater, about the challenges of getting there, being there, and returning to the surface.

Sylvia does not seem very interested in talking about technology. Her reluctance surprises me. As much as anyone else I have interviewed for this book, she has both seen and been part of the technologies that put humans underwater. Her persistence in discussing the environment strikes me as a suggestion that the story of humanity's presence underwater cannot be told without discussing the destruction of the oceans, or, more optimistically, the fight to end the destruction of the oceans, to reverse the damage that has already been done, the rearguard action, the small wins that may—that must—give way to larger victories.

⌒⌒

After talking to Sylvia Earle, I review my interview notes. It turns out that many of the people I spoke with brought up conservation on their own, without prompting.

Don Walsh himself used the word "sustainability" and, with Sylvia Earle and others, has petitioned against certain kinds of ocean development. Example: in 2015, Don wrote a letter on behalf of Ocean Elders to then U.S. secretary of state John Kerry regarding "imminent plans by the five nations bordering the Arctic Ocean to explore and drill for hydrocarbons" at high latitudes. The letter seemed to urge the Obama administration to lead a shift away from fossil fuels. It suggested that oil exploration and production activities in the Arctic "present tremendous ecological dangers." These were not the words of a radical environmentalist, but of a retired submarine commander, a regular contributor to the U.S. Naval Institute's magazine *Proceedings*, a technologist, a man who went to the bottom of the Challenger Deep the year before I was born.

And there was Phil Nuytten's vision for sustainable deep sea mining.

And Graham Hawkes talking of ROVs using very thin fiber-optic control cables, tethers that would be prone to breakage and loss, but, he noted emphatically, that would have to be made from materials that would not harm the seabed. Specifically, he mentioned the possibility of control cables designed to break into pieces that would be no different from the siliceous sand already found on the world's seabeds.

Many of those I interviewed were people who started going underwater to earn a living. They were not born conservationists. They were mid-career converts, which, in the underwater world, are nothing new. Jacques Cousteau himself morphed from spear fisherman to conservationist.

It is, at this point in history, impossible to spend time in the oceans without seeing the degradation, without, as a rational human being, becoming appalled. At times it is difficult to avoid despair. What is out of sight and out of mind for those who stay on the surface is blatantly obvious to those who venture beneath the waves.

The ocean, it is often said, is fragile. Coral reefs, mangroves, kelp beds, and the strange communities surrounding deepwater vents have all been accused of fragility. This is unjust. If it were true, they would be gone. They are, in fact, tough and resilient. They have to be to survive the unrelenting and multifaceted assault from humanity. But to say that they have survived is not to say that they continue to thrive. The ecosystems beneath the waves are damaged and badly wounded. In some cases, the wounds have been mortal. But in other cases, for now and maybe only for now, some underwater ecosystems continue to survive humanity's assault.

That assault includes megatons of sewage, fertilizer runoff, pesticide runoff, sediment, and plastics. It includes pharmaceuticals that leave the bodies of users in their urine, find their way through sewage treatment plants, and are discharged into rivers that flow into oceans. It includes water warmed by power plants and refineries. It includes oil spills, big and small, from wrecked tankers and blown-out wells, but also from the bilges of smaller boats and the leakage from outboard motors and the runoff from parking lots. It includes vessel groundings, ships and barges and even sailboats running into reefs. It includes a warming climate and elevated carbon dioxide levels that change the pH of the water itself. It includes fishing fleets that intentionally and efficiently take target species while unintentionally but with equal efficiency gobbling up everything else that gets in the way, the so-called bycatch that is dumped overboard, much of it dead or dying. It also includes fishing fleets that drag nets and trawls across sea-beds to depths beyond ten thousand feet, sweeping up seafood and bycatch while leaving behind swaths that might as well have been made by submersible Sherman tanks. It includes deafening noises put out by military sonar systems, pile drivers, and air gun arrays used in the search for offshore oil and gas, but also the less deafening but chronic and ubiquitous noises generated by ships, all of which are known to affect whales and other marine life. It includes the introduction of species from one ocean basin to another—say, the Pacific into the Atlantic, or the Atlantic into the Pacific—species such as lionfish and the northern Pacific seastar and European green crab and orange cup coral and, well, a host of others, each changing long-standing communities in often unpredictable and seldom beneficial ways.

Humanity, if it set out to purposely destroy the oceans, might not come up with a more effective strategy than that which is in play today, a strategy of multiple unrelenting impacts, of death by a thousand cuts.

To quote Sylvia Earle: "We need the ocean, and now the ocean needs us."

None of this is to say that there are not laws and policies in place to minimize damage to the oceans. In the United States, for example, there is the National Environmental Policy Act, requiring any activity in the nation's territorial waters that has some level of federal government involvement to undergo a review of possible impacts and how they might be managed. Behind that law there are other laws, nested, working together to protect our oceans. There is the Clean Water Act and the Ocean Dumping Act, for example, both limiting what can be pumped or dumped over the side. Perhaps carrying more power, there is the Endangered Species Act and the Marine Mammal Protection Act. Since it is almost impossible to undertake any activity in the marine environment without at least the possibility of encountering endangered species and marine mammals, both acts have, if fully enforced, wide applicability. All sea turtles, for example, are endangered, and therefore the Endangered Species Act can and has been used to force shrimp fishermen to alter their nets in ways that improve a turtle's chances of survival. And because whales and dolphins are so highly dependent on underwater sound for communication and hunting and navigation, the Marine Mammal Protection Act has been used, under what some believe to be far too limited circumstances, to regulate underwater noise. There is, too, the Marine Debris Research, Prevention, and Reduction Act; the Shore Protection Act; and the Act to Prevent Pollution from Ships. There are marine sanctuaries and marine national parks.

And this is just in the United States. The European Union and most developed nations, along with many less developed nations, have regulations of their own. In addition, there are treaties and international agreements. Enforcement, both in the United States and abroad, may not be what it should be, but rules—at least some rules—are in place.

But none of these rules, sanctuaries, treaties, and agreements have stopped, for example, the incremental industrialization of the Gulf of Mexico stretching close to two hundred miles from shore and to depths below seven thousand feet, and none prevented BP's infamous 2010 spill. None prevent the needless death of hundreds of thousands of dolphins, porpoises, and whales in fishing nets each year. None prevent the existence of 405 documented low-oxygen "dead zones" in various ocean basins, caused by runoff from industrial-scale agriculture. None prevent mining companies from promoting and planning the digging up of mile-deep hydrothermal vent systems, an activity that, for now, is stalled not by environmental protections but by economics.

The world, it turns out, is not so different from Haitian fishermen. Except for this: the Haitian fisherman fish to support an austere and hungry existence, for enough to eat, while the rest of us chase what are, in comparison, luxuries.

⌐

Here in our anchorage in Haiti, in a beautiful harbor at Île-à-Vache, the Island of Cows, the people are so poor that almost no one owns an outboard motor. Instead, they sail in crudely made wooden boats or paddle dugout canoes made, I am told, from the trunks of mango trees. From the cockpit in the back of our boat, my wife and I watch lobster divers working in the early morning sun, diving on a breath of air as they tow their canoes behind them.

I take the opportunity to join a fisherman on his morning rounds. He is in a dugout canoe, with a friend who speaks Haitian Creole and English acting as an interpreter. I follow along behind them in a kayak.

The fisherman's dugout canoe, like almost all dugout canoes, leaks, and after every ten or twenty paddle strokes, he scoops water from the canoe using a plastic bailer made from what was once a soft drink bottle. His transition from paddling to bailing to paddling is seamless, and despite the extra effort, I have trouble keeping up with him and the interpreter.

The fisherman wears torn shorts and a ripped T-shirt that in most places would long ago have been converted into rags. He has two fins, one blue and one black, both with broken blades, making them half fins. His snorkel, too, is broken, so when he is facedown in the water, its top barely reaches the surface.

We paddle a mile from shore. The interpreter stays in the canoe while the fisherman and I drop into the water. Turtle grass covers the bottom. I see no coral or rock. The fisherman immediately dives to ten or twelve feet, hovering in the water column. He stays there fifteen or twenty seconds before descending to the bottom. He reaches into the seagrass to retrieve a tiny lobster. From head to tail, the lobster is less than six inches long.

He dives repeatedly, typically for about thirty seconds. His time on the surface between dives is maybe forty seconds. Up and down, up and down, up and down. Every third or fourth dive, on average, he comes up with another tiny lobster or a juvenile conch. On one dive he retrieves a black sea cucumber.

I know this sort of thing is commonplace, but I am troubled by what I see. Later, through the interpreter, I ask if the fisherman understands that these lobsters and conchs are too small to take. He tells me that, yes, he knows, but he also has to earn a living.

I do not know the fisherman well, but his knowledge of languages seems restricted to Haitian Creole, a few words of French, and a couple of words of English. I do not think he can read. He does not own land. What options does he have in life? He can take the tiny lobsters and conchs and the occasional sea cucumber, or go hungry. Or he could beg. Or, maybe, he could steal.

As it turns out, it is not quite right to say that he eats his seed stock. His catch is too valuable to eat. He sells it. With the money he buys beans and rice and cassava and the mealy black-eyed peas that, at the moment, seem to be abundant in the village. He does not know where the tiny lobsters and conchs and sea cucumbers are eaten. That is not his concern. But he thinks they are shipped to the city—to Port-au-Prince—to be served in restaurants.

The fisherman, I should mention, has the face of a fifty-year-old, but in fact he is in his thirties. He is also as lean as a street dog.

He has occasionally seen turtles, which are mostly gone from Haiti, but once in a long while he crosses paths with one, or a friend will bring one up in a net. When the odd turtle is captured, it, too, is sold as food. But because there is not always a ready market for sea turtle meat, even in Haiti, he has eaten it himself. It is, as translated by the interpreter, "tasty."

Dolphins, it turns out, are even rarer than turtles in this part of Haiti. But, my interpreter tells me, even tastier.

When we paddle back toward shore, the diver is noticeably worn out. His strokes are slower than they were on the way out, and he stops twice to rest. The diving, up and down and up and down, fueled by nothing more than beans and rice, is exhausting.

I see a tuna jump, and then another, and another, all right out in front of the village. Their silver-and-blue sides sparkle in the sun. They surprise me, occurring here in this desert of over-harvest. They give me a spark of hope. Maybe the situation is not as bad as I think.

An hour later I watch four fishermen stretch a gill net across the bay. The net, in its entirety, is at least four hundred feet long, designed in such a way as to entangle fish rather than to pen them in. The men work from dugout canoes, and once the net is stretched out, they strap on old diving masks that elsewhere would be consigned to the rubbish bin.

Rather than pulling the net out of the water to empty it, they gulp air and dive, swimming along the net, removing entangled fish, throwing them into the canoes. The larger fish, the tuna, have to be killed to prevent them from flopping back into the water, and the men do so by repeatedly bashing their heads against the sides of the dugouts. The method makes a *thump, thump, thump* drumming against the wood. Red blood flows into the water. The smaller fish, many less than six inches in length, are simply tossed into the canoes, where they flip and flop, shimmering in the sunlight.

I turn away, concerned that the fishermen might, as unlikely as it may be, find a turtle in their net. If such a thing did happen, I am not sure what I would do. Perhaps I would buy the animal and release it. But it seems unlikely that the animal could be released in this hungry village without controversy, without leaving the impression that I care more about turtles than about people.

That evening, back aboard our boat, a floating enclave of relative wealth with its own solar panels, water supply, well-stocked refrigerator, and internet connection, I succumb to a minor bout of depression. I have spent the last year diving and thinking about diving, considering the innovative spirit that has let humans penetrate the depths, but now I have been confronted with what is happening in Haiti, and by extension what I know is going on elsewhere on scales large and small. I think of myself as a problem solver, as someone reasonably capable of getting things done, but here it seems to me that there is nothing I can do, nothing meaningful to stem the tide.

It is not just desperate Haitians who leave me depressed. It is every fisherman harvesting seed stock. It is oil companies developing offshore oil fields to produce oil that the world could live without. It is farmers dumping fertilizers and pesticides on their crops because, if they ignore the downstream costs, it increases the bottom line. It is everyone who buys cell phones and computers that need rare metals, creating a market that will inevitably drive miners into deep water.

To distract myself, I take a few hours to catch up on the news, which aboard a wandering sailboat is not always readily available. But here in Haiti, oddly enough, I have good internet access.

I read, among other things, that Belize has pledged to ban single-use plastics, such as shopping bags, picnic forks, and straws. I take heart in this knowledge, but I immediately wonder why such laws can be passed in developing nations but not in the United States. And I wonder if the law could somehow be extended to keep smaller fragments of plastic out of the seas, the sort that come from washing synthetic clothes and then draining the wash water into streams and rivers that lead to oceans, the fragments that, it turns out, are the same size as the plankton eaten by small fish and filtered from the water by corals and sponges.

I read, too, a story about Elon Musk. One of his companies, SpaceX, launched a car from one of his other companies, Tesla, into space. It was not just any car. It was a Tesla convertible roadster, a beautiful electrically powered red convertible that had been Musk's personal vehicle. In the driver's seat sat a mannequin in a space suit with its right hand on the steering wheel and its left elbow propped on the edge of the door. The car also carried a copy of *The Hitchhiker's Guide to the Galaxy* and the Isaac Asimov Foundation trilogy.

After spending the morning with a diving fisherman, after worrying about the unlikely capture of a sea turtle, I am outraged and sickened by the Musk story. But I am also pleased. He has accomplished something here. Sending a sports car into space seems pointless, a waste of resources, but it has generated excitement and interest in space and technology and innovation.

And with that thought, I wonder, for a moment, what it would take to get Musk and other billionaires to step up their game for the world's oceans. There is, to be fair to Musk, his involvement with the XPRIZE Foundation, with, for example, its Wendy Schmidt Ocean Health prize, but that is not quite the same as the wholehearted and obsessive involvement that it takes to, for example, become a billionaire, to start an endeavor like SpaceX, to build beautiful fully electric roadsters.

And then I realize that the issues here are not the sort that can wait for the attention of a billionaire. I must act and do what I can in my own life to avoid adding to the onslaught against the oceans.

Divers, Sylvia Earle suggested, are ambassadors, and I choose to interpret her "divers" as all-inclusive, as anyone who ventures beneath the waves.

"Divers," she told me, "have the opportunity and the responsibility to share with others." By extension, they must do whatever is in their power to secure the future of the oceans.

And that is, I realize, the very least I can do.

EPILOGUE

In Haiti and later in the Netherlands Antilles, as this book made its way through the publication process, I reviewed my notes, references, and draft chapters. In doing so, I sometimes felt as if I were talking to my younger self—my deep sea diver youthful self, with his full head of hair and his attitude of indestructibility—about the past and future of the underwater world. And during that conversation, well after the deadline had passed for making significant changes to the book's text, two events of special relevance unfolded.

First, the remains of the Argentine submarine that had been lost with all hands in 2017 was located by Ocean Infinity. The discovery came through the company's ongoing use of HUGIN AUVs, the same robots (or similar ones) that had been deployed in what has been, to date, the unsuccessful search for Malaysian Airlines Flight MH370, described in chapter 6. Although locating the wreckage of the submarine will do little to soothe those who lost loved ones, it does validate Ocean Infinity's faith in the utility of AUVs working with substantial independence, which also suggests that Don Walsh's vision of a world in which AUVs do "all the heavy lifting" of deep ocean exploration is upon us.

Second, another private sector enterprise intends to send a manned submersible to the deepest points in the Atlantic, Pacific, Southern, Indian, and Arctic Oceans. The Five Deeps Expedition, led by billionaire adventurer Victor Vescovo, is using a titanium-hulled submersible built in secret by the remarkably innovative company Triton Submarines of Sebastian, Florida, for what Don Walsh has called "one of the most ambitious exploration expeditions of the century." From Patrick Lahey, co-founder and president of Triton: "We worked for over three years to develop the design and required technology. The submersible, together with the unique and specially equipped support vessel DSSV *Pressure Drop,* promise to change our relationship with the deep ocean in a profound and fundamental way." And this, from Vescovo, following completion of the expedition's first major dive, which took him and his submersible to the bottom of the Puerto Rico Trench at a depth of 27,480 feet: "It was an extraordinary dive, and I am so proud of our international team for making it possible over the last few years. It felt great to get to the true bottom of the Atlantic Ocean for the first time in history and to prove the technical capabilities of this diving system, which we believe is now the deepest operational one in the world. We are really looking forward to continuing to the other dive sites, and continuing our technical and scientific goals."

The team's second landmark dive, to 24,388 feet in the Southern Ocean's South Sandwich Trench, was completed in February 2019, and their third landmark dive, to 23,596 feet in the Indian Ocean's Java Trench, was completed in April 2019. If all goes well—and needless to say the plans of the Five Deeps Expedition are far from risk-free—Vescovo and his team will visit the Challenger Deep about halfway through 2019, around the same time this book will be released.

The history of humanity's presence beneath the waves advances. And while robots may do the heavy lifting, a key role remains for manned expeditions to contribute not only technological advances and useful data but also that most important of all commodities: inspiration. Victor Vescovo and his team may offer exactly the kind of inspiration needed to spur further understanding and, by extension, conservation of our oceans.

ACKNOWLEDGMENTS

Writers spend way too much time sitting in front of keyboards and glowing screens, but they do not work alone. This is especially true of nonfiction writers. So a few words of thanks, as always, are in order.

First and foremost, thanks to everyone who agreed to be interviewed for this book or who provided written material. Most of those whose interviews or written documents were incorporated into the book are named in the text, the endnotes, or both. A few of those who are not named and in some cases not directly quoted but whose thoughts contributed to this book are Tyler Whitmore, Greg Murphy, Vince Ferris, Clarice Cote, Eduardo Moreno, Brodie Kraft, Miles Laster, Brian Hoover, Elisa Miller, Erik Thorbjorsen, Henry Liyanto, Andreas Widy, Chris Combs, Craig Cooper, Ted Curley, the staff of Triton Submarines in Florida, and the staff of U-Boat Worx in the Netherlands. To all of you, thank you for your time and for trusting a stranger with your thoughts.

I owe a special thanks to the U.S. Navy. Various personnel in the Navy responded to a Freedom of Information Act request frankly and enthusiastically. More important, the Navy approved access to two submarine bases and its Experimental Diving Unit, allowing me to talk to various active duty divers, submariners, and civilian employees, all without intrusive supervision or interference. The Navy also provided valuable comments on a draft version of *In Oceans Deep*. I owe a special thanks to Lieutenant Jennifer Jewell, Ph.D., U.S. Navy research psychologist and Experimental Diving Unit public affairs officer.

Thanks to Jean Samuel Altema of Île-à-Vache, Haiti, for arranging conversations and outings with local fishermen, as well as interpreting from Creole to English.

Many thanks to John Parsley, who helped with the development of the proposal and outline that became *In Oceans Deep*, and many thanks to Philip Marino, my editor at Little, Brown. Copyeditor Barbara Jatkola offered diplomatically worded and encouraging corrections to boneheaded grammatical and even factual errors, and Gregg Kulick's cover design captures the spirit of the book. Many others at Little, Brown, often working behind the scenes in ways that an author will never understand, helped with this book, including Betsy Uhrig, Anna Goodlett, and Reagan Arthur.

A number of friendly readers reviewed all or parts of an early draft of *In Oceans Deep*. These included my wife and diving partner, Lisanne Aerts; writers Kris Farmen, Lori Townsend, and John Clarke; and Mark "Tex" Rogers, Don Walsh, and Bob Christ.

To all of you and many others, I hope *In Oceans Deep* more than justifies the time you spent talking to me, providing references, and reading draft manuscripts.

NOTES

CHAPTER 1: DESCENT

The *Trieste* story is available from many sources, including official government reports and various narratives summarizing those reports. Jacques Piccard, with coauthor Robert Dietz, offered his version of *Trieste*'s history, including its famous dive, in *Seven Miles Down* (1961, Putnam, New York). According to Don Walsh, Piccard did not tell him or their close associates about the book until it was finished, perhaps reflecting Piccard's feeling that, as a Swiss citizen and a civilian, he was an outsider rather than an integral part of the Navy's *Trieste* team, even though it was Piccard and his father who developed *Trieste* and brought it to the Navy. Piccard, during the dives leading up to the Challenger Deep dive and during the Challenger Deep dive itself, was under contract with the U.S. Navy but was not a direct employee. To my knowledge, *Seven Miles Down* is no longer in print, and it does not seem to be available electronically. I found a used copy that once belonged to the South Carolina State Library Board. In the back of my copy, two of the old-fashioned check-out cards, familiar to library users before the twenty-first century, were untouched, sadly suggesting that the book had never left the shelves until it was sold. And, surprisingly, about halfway through the book I realized that I had read it long ago, likely before my teenage years, although this memory cannot be confirmed and may not be accurate. In any case, most of my direct quotes of conversations reported as occurring during *Trieste*'s Challenger Deep expedition come from *Seven Miles Down*. A few of the quotes come from U.S. Navy documents, including old media releases and reports.

One of my favorite Navy sources was a yellowing copy of *The Bathy-scaphe TRIESTE, Technological and Operational Aspects, 1958–1961* (Research Report 1096, U.S. Navy Electronics Laboratory, San Diego, July 27, 1962), written by Lieutenant Don Walsh, who would have been, at the time he wrote the report, just turning thirty years old.

Don Walsh explained to me that he was the officer in charge during the *Trieste* mission, a designation that, in Navy jargon, is used aboard vessels that are not Navy commissioned. He was, in essence, the captain. Lieutenant Larry Shumaker was the assistant officer in charge.

My 2016 book, *And Soon I Heard a Roaring Wind* (Little, Brown, New York), describes how my wife and I crossed the Gulf of Mexico and sailed to Guatemala while also telling the story of weather forecasting, starting with primitive efforts in the 1800s to today's complex and far more accurate efforts that rely on mathematics and supercomputers. Before that crossing, the longest passage we had made was a single overnight sail near the Texas coast. In that book, I explain that we were sailing to live closer to the environment, and that statement holds true today. Despite the inconveniences and discomforts that come with living on a small cruising sailboat, we remain aboard, sailing and learning about the oceans, and we intend to continue this lifestyle as long as our health allows.

William Beebe's story is now all but forgotten, but Walsh and Piccard, in 1960, would have been well aware of Beebe's dives, and almost certainly would have read his popular book *Half Mile Down* (1934, Harcourt, Brace, New York).

Auguste Piccard was far more than a mere inventor. He was a scientist in search of data, and to find that data he sometimes had to invent new means of transport. For example, his high-altitude balloon adventures were motivated by his desire to measure cosmic radiation (which could not be reliably measured at low altitudes or even from high mountains). Don Walsh suggested that I include more information in the text about Piccard's genius, but I thought the digression would distract readers from the main story.

Scuba tanks seldom explode, in part because their valves have a "burst disk"—that is, a small metal disk designed to release the tank's air (or other gases) at pressures well below those associated with catastrophic failure.

The gasoline used in *Trieste* was aviation gasoline, or avgas, intended for use in aircraft with spark-ignited internal combustion engines. Avgas is essentially leaded gasoline; commonly used automobile gasoline no longer contains lead.

Metallic lithium and liquid ammonia were also considered for use in *Trieste*'s float, but gasoline had several advantages, not the least of which was that it was inexpensive and available throughout the world. It is not strictly correct to say that liquids are incompressible. Gasoline's approximate bulk modulus (a measure of compressibility) is 1.3×10^9 Pa. Water, also often said to be incompressible, has a bulk modulus of about 2.15×10^9 Pa. At the bottom of the Challenger Deep, both gasoline and water are somewhat compressed and therefore denser than they would be at the surface, but for most practical purposes they are considered incompressible; as an engineer or a physicist might say, "On first approximation they are incompressible."

At least one source suggests that *Trieste* relied on nine tons of iron shot as ballast. Don Walsh corrected this error, explaining that *Trieste* carried sixteen tons of shot—eight tons in each of two separate ballast tubs. Errors and misconceptions about *Trieste* are all too common in the literature, but I have confidence that Don remembers how many tons of shot were used.

A note regarding the spelling of "bathyscaphe" is in order. The word is from the French *bathyscaphe*. *Trieste*'s Italian builders would have called her a *bathyscapho*. In English, it is usually spelled "bathyscaphe," although none other than Don Walsh favors its alternative spelling, "bathyscaph."

Trieste went through significant modifications over time, and ultimately more than one bathyscaphe was named *Trieste*. The *Trieste* used by Walsh and Piccard—or at least the parts of her that were not replaced over her many years of service—is on display in the National Museum of the United States Navy in Washington, D.C. *Trieste II,* built later but in most ways similar to the first *Trieste,* is on display in the U.S. Naval Undersea Museum in Keyport, Washington.

Chris Wright's wonderful *No More Worlds to Conquer: Sixteen People Who Defined Their Time—and What They Did Next* (2015, HarperCollins, London) starts with a chapter on Don Walsh, describing, among other things, the challenge of reaching Don's house in rural Oregon. Another chapter in the book features Joe Kittinger, who jumped from a gondola suspended beneath a helium balloon at an altitude of just over nineteen miles in 1960. (Of course, he wore a parachute. He was not, after all, suicidal.) Astronauts are also covered. Another chapter—and I compliment the author's sense of story as well as his sense of humor for including this chapter—covers Gloria Gaynor, who sang the 1970s disco hit "I Will Survive."

From an email Don Walsh sent me regarding involving scientists in decisions about *Trieste:* "In 1957 the Office of Naval Research contracted with the Piccards to do a series of test dives at Capri. Dives were offered to oceanographers from different disciplines (e.g., biology, geology, acoustics, etc.). They were asked to evaluate use of this type of platform for their specific disciplines. At the end of the seventeen dive series, a 'committee of the whole' advised the Navy (ONR) that they should acquire *Trieste.* This was done by ONR in January 1959 with *Trieste* arriving at the Navy Electronics Laboratory in San Diego in October of that year."

The biography that records Lord Nelson's quip regarding his right to be blind is *The Life of Horatio Lord Nelson,* by Robert Southey, published in 1813 in London and available online via Project Gutenberg (http://www.gutenberg.org/ebooks/947). Although some scholars question the veracity of the quote, it has become an entrenched aspect of Nelson's history. The account of Nelson's behavior during the battle probably gives us the well-known expression "turn a blind eye." Another excellent biography is *The Life of Admiral Lord Nelson, K.B., from His*

Lordship's Manuscripts, by James Stanier Clarke and John M'Arthur, published in 1810 in London. The book remains available electronically and in print form.

To modern readers, intentionally dumping gasoline into the sea may seem reprehensible, but in 1960 unwanted fuel was routinely dumped overboard. Although it would not be an acceptable practice today, it is worth remembering that gasoline tends to volatilize quickly from the sea surface, especially in the tropics, and it is potentially less harmful than heavier hydrocarbons such as diesel, bunker fuel, and motor oil. It may also be worth remembering that despite many laws, bilge water contaminated with hydrocarbons continues to be pumped over the side by ships, oil leaks incessantly not only from pipelines but also from natural seeps, and tankers run aground and spill oil all too often.

The loss of consciousness of Austrian free diver Herbert Nitsch during his 2012 record-breaking (and astonishing) breath-hold dive to 831 feet may have been caused by nitrogen narcosis, but other factors may also have been at play. The physiology of extreme-depth free diving is not well understood. For obvious reasons, measurements are difficult to collect during breath-hold dives to great depths.

Egyptian diver Ahmed Gabr holds the world depth record for scuba diving. See chapter 4 for more.

For a wonderful book about jellyfish, pick up Juli Berwald's *Spineless: The Science of Jellyfish and the Art of Growing a Backbone* (2017, Riverhead Books, New York).

The question "What kid wants to grow up to be a robot?" (and variations of it) is often uttered in exploration circles, but it seems to have originated with a response given by James Cameron when asked about the value of sending human beings to great depths (and, by extension, into space).

Through the Freedom of Information Act (FOIA), I tried to determine more about the orders limiting *Trieste*'s dives to 20,000 feet. The Navy, after initially asking for funds to pay for my FOIA request, acknowledged that the request fell into a category that could be funded by the Navy itself. Days went by, and then weeks. My request was passed between departments. Ultimately, I received several responses, each expressing what appeared to be genuine interest in the question, but all essentially saying that no one was sure of the origin of the order. One added that the Navy, which now justifiably takes great pride in the history of *Trieste,* would be very interested in anything I might uncover about the order, but as of this writing my efforts have led to nothing but dead ends. The U.S. Navy is many things, including, as anyone who has served is well aware, a huge and relentless bureaucracy.

CHAPTER 2: ON A BREATH

The United Nations Educational, Scientific and Cultural Organization (UNESCO) estimates that three million shipwrecks litter the ocean floor. The wreck of the *Oliver* was discovered—actually, rediscovered, since her sinking would have been well known to the people of Utila in 1803—in 1971 by Chris Talbert and Gunter Kordovsky. Gunter told me that the wreck was buried by and overgrown with coral and sponges when he first saw it. With other divers, he excavated the wreck and recovered anchors, a small cannon, and other artifacts. Today, it is a marked dive site just off the shore of Utila, but it is not one of the island's more popular dive sites. Apparently, most visiting divers see only what can be seen—the pile of ballast stones and a piece of the hull lying on the sand—rather than what has to be imagined. That is, they do not see the lives of sailors working there in 1803, the striking of the reef, the unsuccessful fight to save the vessel, the unknown fate of Captain Hood and his crew, or any of the work that went into uncovering the wreck in the 1970s, when scuba diving was, on Utila, still something of a novelty.

My free diving instructor was Mark "Tex" Rogers, who owns and operates Freedive Utila. This school manages to attract and retain dedicated instructors who know how to match their teaching techniques to the needs of different students, and while there is an inevitable focus on depth, instructors frequently repeat one of Mark's rules of free diving: "Never do anything hard."

Oxygen levels in the blood are sometimes estimated using a pulse oximeter clipped to a fingertip. Changes in light as it passes through the fingertip indicate the amount of oxygen in the blood. This, however, does not offer a good estimate of oxygen saturation in the brain during breath-hold dives, because circulation to the extremities is restricted.

The poet Samuel Taylor Coleridge attended the third meeting of the British Association for the Advancement of Science on June 24, 1833, in Cambridge. "You must stop calling yourselves natural philosophers," he proclaimed, apparently irritated by what he may have seen as a certain arrogance, a certain cockiness, that allowed those who collected rocks, dirtied their hands with experiments, and spent time at sea to consider themselves philosophers of any kind. A lesser-known man, William Whewell, caught the attention of the audience. "If 'philosophers' is taken to be too wide and lofty a term," he said, "then, by analogy with 'artist,' we may form 'scientist.'" And so scientists they and their successors became.

Records, in free diving as in other endeavors, are ephemeral. They are also associated with closely defined rules. At the time of this writing, the record for a breath-hold dive after breathing ordinary air was 10 minutes and 49 seconds, but millions of television viewers in 2008 watched a breath-hold dive of 17 minutes and 4 seconds on an episode of *The Oprah Winfrey Show*. The difference? Oprah's diver breathed pure oxygen before the dive. Not long after the episode aired, the "static apnea on oxygen" record, as it is called, was extended to 22 minutes and 22 seconds.

The yoga manual referenced in the text is *Asana Pranayama Mudra Bandha*, published by the Bihar School of Yoga in Munger, India. I have no idea what the authors would think about the use of their techniques in free diving, but it seems like the sort of thing that might meet with their approval. My guess is that the teachers at the Bihar School would be very good free divers.

Diagnosing lung and throat squeezes in free divers by examining the victims' spit may be useful in the field (that is, floating on the surface), but it is far from certain. Lung squeezes can be deadly, and the possibility of a lung squeeze should be taken very seriously. If in doubt, consult a qualified physician.

Captain Spratt wrote of Greek sponge divers in his two-volume *Travels and Researches in Crete,* published by John van Voorst in London in 1865 and available today online at https://archive.org/details/travels andresea01spragoog/page/n3. "Marvelous as it may appear," he wrote, "yet such is the fact, that man's power of versatility is so great that he can be brought to endure suspended respiration and to sustain the pressures experienced during a dive to depths of twenty or thirty fathoms; nay, I have been assured that in some few instances he has been known to reach forty fathoms without the aid of any diver apparatus save a flat stone carried in his hands to facilitate his descent."

References about Hatzis's dive are confusing and not altogether reliable. What I have presented is based on a combination of what I believe to be the most reliable sources. Of these, by far the most thoroughly researched account comes from *Freediving: The Story of Stathis Hatzis and the History of Sponge-Fishing Freedivers in Greece,* by Yannis Detorakis (2014, published by the author). In the preface, the author writes, "It has already been twenty years since that day when I decided, having only three or four newspaper and magazine articles in my hands mentioning a man called George Hatzistathis, to risk my entire fortune at the time as a young professional, on an extensive research to discover the true story of Stathis Hatzis."

PADI leadership, despite several exchanges by email, did not respond to questions I provided about free diving safety.

Per Scholander's 1940 monograph, *Experimental Investigations on the Respiratory Function in Diving Mammals and Birds* (Hvalrådets Skrifter, no. 22), is without doubt one of the most important scientific papers ever published about diving physiology, but Scholander's contributions went far beyond diving physiology. For example, he also published on fluid movements in plants and on cold adaptations in animals. He was among the modern pioneers of science in Alaska and is often credited, along with his collaborator (and father-in-law) Laurence Irving, with establishing the research laboratory in the northernmost large community in America—Utqiagvik (previously Barrow), Alaska. He was instrumental in the design and construction of the famous research vessel *Alpha Helix,* and he once organized and participated in an unauthorized but successful military rescue mission. But for all his accomplishments, he was known for his warm and enthusiastic personality and his sense of humor. For example, after describing how he believed dolphins could so efficiently bow ride in front of ships, he wrote, "This, I believe, is the way dolphins ride the bow wave, and if it is not, they should try." For a short biography of Scholander, see https://www.nap.edu/read/897 /chapter/15#397. Alternatively, read the still available but out-of-print *Enjoying a Life in Science: The Autobiography of P. F. Scholander* (1990, University of Alaska Press, Fairbanks).

Researchers continue to tag seals and whales, often with surprising results. The dive depths for ringed seals described in the text comes from animals tagged in Alaska in a project co-managed by biologist Andrew VonDuyke of the North Slope Borough's Department of Wildlife Management in Alaska. Tagging is done with the assistance of Inupiat (Eskimo) hunters.

The quote from the *Journal of Applied Physiology* is from the 2009 article "The Physiology and Pathophysiology of Human Breath-Hold Diving," by Peter Lindholm and Claes Lundgrend (vol. 106, no. 1, pp. 284–92). Similar review articles are available in many academic journals.

The story of Nick Mevoli is well known in the free diving world, and it has been described in many sources. Among the most well researched of these is Adam Skolnick's *One Breath: Freediving, Death, and the Quest to Shatter Human Limits* (2015, Crown, New York). Skolnick's book also offers insights into the realities of competitive free diving—an international world full of gossip, with an ethos that both eschews and is awash in egotism, and where people are injured and killed with alarming frequency.

I am not an expert on Ben Franklin, but I am a fan. Based on my knowledge of how he lived, I believe that if he were alive today as a reasonably young man, he would be a free diver. He was, at a time when swimming was shunned by most, an avid swimmer, and clearly he was an inveterate experimenter.

Halley's wonderful essay on his diving bell is available online at https://ia802905 .us.archive.org/17/items/philtrans05114087/05114087.pdf, and an illustration of his diving bell at work can be found online at Wikimedia and elsewhere by searching for "Halley's diving bell." For additional information on diving bells and their evolution, see Bryce Emley's condensed and admirably readable "How the Diving Bell Opened the Ocean's Depths," published in the March 23, 2017, edition of *The Atlantic*. Halley's bell also appears in historical fiction, as I recently discovered to my delight while reading *Treason's Harbor* (Norton, 1989), volume 9 of Patrick O'Brian's Aubrey-Maturin series, known for its accuracy with regard to nineteenth-century naval operations and ship handling..

CHAPTER 3: UNDER PRESSURE

The officially recognized free diving record in 1971 was held by Enzo Maiorca. In the years before and after, he traded the record-holding title repeatedly with Jacques Mayol. Their competitive relationship was fictionalized in the 1988 film *The Big Blue*. Mayol committed suicide in 2001, at age seventy-four. Maiorca entered politics and lived to be eighty-five years old.

Some sources suggest that the U.S. Navy heavily publicized the installation of DSRVs on submarines as a counterintelligence measure designed to prevent the Soviets from looking too closely at the diving system disguised as a DSRV on the USS *Halibut*. However, DSRVs had strong public relations value as well. American citizens wanted to feel like their Navy cared about sailors. In addition, even though DSRVs could save lives only under specific circumstances, they undoubtedly brought some level of comfort to submariners. A DSRV is in some ways similar to the airbag system in a car: it will not save your life in every accident, but it sure is comforting to know it is there.

Operation Ivy Bells has been described in numerous articles (many with questionable facts) and in the excellent 1998 book *Blind Man's Bluff: The Untold Story of American Submarine Espionage,* by Sherry Sontag and Christopher Drew (Perseus Books, New York). There is also a self-published 2014 first-person novel, *Operation Ivy Bells,* by Robert G. Williscroft (Starman Press, Carson City, NV)—an engaging read with larger-than-life characters that brings the reader into the guts of Ivy Bells, but that may stretch certain realities to enhance the story.

In 1834, Victor Junod described the symptoms of nitrogen narcosis. "The functions of the brain are activated, imagination is lively, thoughts have a peculiar charm and, in some persons, symptoms of intoxication are present," he wrote. For some five decades after Junod's work, narcosis was attributed to changes in blood flow caused by changes in pressure; somehow, researchers incorrectly suggested, the blood is pushed along by the increased pressure. In 1899 and 1901, Hans Meyer and Charles Overton separately recognized the link between fat solubility and narcotic effects. For example, highly fat-soluble volatile substances such as chloroform are also highly narcotic even at low pressures. By 1939, the narcotic potential of other gases under pressure was demonstrated. Xenon, it turns out, is so narcotic that it could be (and sometimes is) used as an anesthetic at one atmosphere of pressure. It is odorless, harmless to the environment, nonexplosive, and safe. But for its cost, it might be a common anesthetic today. The Meyer-Overton hypothesis stands, more or less intact, although the underlying mechanisms of narcosis remain only partly understood.

Alphonse Jaminet's report on what came to be known as decompression sickness is available at https://www.forgottenbooks.com/en and elsewhere online. Its rather wonderful full title is *Physical Effects of Compressed Air; and of the Causes of Pathological Symptoms Produced on Man, by Increased Atmospheric Pressure Employed for the Sinking of Piers, in the Construction of the Illinois and St. Louis Bridge over the Mississippi River at St. Louis Missouri* (1871, R.T.A. Ennis, St. Louis). "Having no pretensions as a writer," he charmingly begins the report, "I ask the indulgence of my readers in offering this result of my thoughts and observations."

Martin Goodman provides a good biography of John Scott Haldane in *Suffer and Survive: Gas Attacks, Miners' Canaries, Spacesuits and the Bends; The Extreme Life of Dr J. S. Haldane* (2007, Simon & Schuster, London). The biography includes an account of the sealed-box experiment, as well as accounts of Haldane's dangerous descents into mines and sewers.

John Scott Haldane, A. E. Boycott, and G.C.C. Damant published their 101-page 1908 paper, "The Prevention of Compressed-Air Illness," in the *Journal of Hygiene*. While that journal is not to be found on every diver's bookshelf, the article itself, now well over a century old, deserves a wider readership. Fortunately, it is readily available online and for the most part does not require scientific training to be understood and enjoyed. It includes Haldane's tables as well as descriptions of test dives, including dives by his then seventeen-year-old son. In 2006, Alexander von Lünen, writing in *Wilderness and Environmental Medicine* (vol. 17, pp. 195–96), offered a short paper called "Goats and Gases: 'The Prevention of Compressed Air Illness' by Haldane et al—*A Commentary*," reminding readers that Haldane never claimed (nor deserved) sole credit for the development of decompression tables and decompression theory. Haldane's most important contribution, according to von Lünen, was the development of the so-called 2:1 ratio, the idea that a diver could safely reduce exposure to pressure by one-half without developing serious decompression sickness. Other important contributions outlined in Haldane's paper, including the use of staged decompression and the tables themselves, came with a large measure of help from his collaborators and earlier researchers, including Paul Bert. Nevertheless, Haldane is thought of today as the father of decompression tables and, by extension, the father of the algorithms used in decompression computers. Without the work usually credited to Haldane, basic scuba diving, as well as the technical, commercial, and military diving that are routinely practiced today, would be deadly.

CHAPTER 4: SATURATED

Patents can be found and reviewed easily online at http://patft.uspto.gov/.

The story of Max Nohl's dive to 420 feet has been told many times. One excellent summary is available at http://aquaticcommons.org/14994/1 /Historical_Diver_7_1996.pdf. A good source for a more general history of decompression is the 1998 book *The Bends: Compressed Air in the History of Science, Diving, and Engineering,* by John L. Phillips (Yale University Press, New Haven, CT).

Modern Mechanics published the article describing how to build a diving helmet from a hot water heater in January 1932. Later that year, the magazine became *Modern Mechanix*. It was similar in many ways to today's *Popular Mechanics*. Enough helmets were made from hot water heaters that some survive today. They can be seen in private collections and, occasionally, for sale, usually with a warning suggesting that they not be used in the water. See https://www.youtube.com /watch?v=o9J4EccBvr8 for an interesting (but sexist) newsreel dated February 18, 1935, showing these helmets in use in Venice, California.

Leon Lyons's *Helmets of the Deep* provides an extensive illustrated history of diving helmets. Although the book is out of print and difficult to find, Lyons is working on a revised and expanded version that he hopes to release soon.

George Bond and his colleagues, like thousands of other scientists employed by the U.S. government, wrote detailed reports about their research. Bond's reports are succinct, clear, and interesting. The report on his animal experiments quoted in the text is *Prolonged Exposure of Animals to Pressurized Normal and Synthetic Atmospheres,* by Robert D. Workman, George F. Bond, and Walter F. Mazzone (U.S. Naval Medical Research Laboratory Report No. 374, Bureau of Medicine and Surgery, Navy Department Research Project MR005.14-3100-3.02, January 26, 1962).

Bob Barth's 2000 memoir, *Sea Dwellers: The Humor, Drama and Tragedy of the U.S. Navy SEALAB Programs* (Doyle Publishing, Houston), provides a valuable firsthand account not only of Sealab but also of what it was to be a Navy diver in the 1950s and 1960s. The book is out of print, but Barth purchased all remaining copies and donated them to the Man in the Sea Museum in Panama City Beach, Florida (at least so I was told at the museum). As of this writing, copies were still available through the museum at the regular retail price of $16.95, with profits supporting the museum. Used copies are also available online at prices ranging upwards of $80.

Bond's summary words about his six years of animal and human experiments and humanity's readiness to station divers on the continental shelf come from his September 1964 article, "New Developments in High Pressure Living," published in the academic journal *Archives of Environmental Health* (vol. 9, pp. 310–14).

Bond's words about his team's commitment to working in an "orderly, carefully documented, scientific manner" and related quotations scattered through the text come from the 1993 account *Papa Topside: The Sealab Chronicles of Capt. George F. Bond, USN,* edited by Helen A. Siiteri and published by the Naval Institute Press (New York). *Papa Topside* is based on various papers, speeches, and reports written by Bond before his death, in 1983.

The Navy has used and continues to use various kinds of rebreathers. The rebreather referred to in relation to Sealab was the Mark VI rebreather. I did not name it in the text because I did not want to confuse readers unfamiliar with Navy nomenclature, especially those who might, at the mere mention of "Mark," immediately think of the famous (within diving circles) Mark V helmets.

Bond's words about "personal idiosyncrasies" come from the fascinating and detailed *Project Sealab Summary Report: An Experimental Eleven-Day Undersea Saturation Dive at 193 Feet,* by H. A. O'Neal, G. F. Bond, R. E. Lanphear, and T. Odum, dated June 14, 1965, and available online at http://www.dtic.mil/dtic/tr/fulltext/u2/618199.pdf. The report includes a short preface by Rear Admiral J. K. Leydon. "Project Sealab I was the U.S. Navy's first step into inner space," wrote Leydon, in what can only be thought of as hyperbole, given that the U.S. Navy had in fact made many first steps into inner space before Sealab I became possible, though it is nice to see the admiral's apparent exuberance. Leydon also wrote of "man's total conquest of the oceans" and of resources in the oceans: "No one discounts the strategic military importance to be derived from a more thorough understanding and complete utilization of the sea. However, the undersea research efforts of the Navy are helping to point out the vast resources within the sea which are available to the benefit of all mankind." It is also worth noting that Leydon's choice of words came years before Neil Armstrong would use similar wording ("step" and "mankind") when he became the first human to stand on the surface of the moon, in 1969.

You can listen to the very entertaining conversation between Carpenter and President Johnson at https://www.youtube.com/watch?v =Gg0pMbc7Opk.

Bond assessed divers reporting symptoms of decompression sickness during the final ascent from Sealab II and dismissed all of them as mere muscle aches. His diagnoses were based on observed symptoms and the absence of progressive worsening as the ascent continued. In one case, three divers had spent five hours sitting over a card table playing cribbage before reporting pain in their lower thighs. Bond, with his typical humor, diagnosed "cribbaticus" and ordered the men to sleep it off. Later X-rays of Carpenter, one of the divers who had reported symptoms during the Sealab II ascent, showed bone lesions above his knees, where he had complained of pain. Aseptic bone necrosis, known to occur in caisson workers, also became associated with saturation diving.

The report describing Sealab II's interest in testing the ability of humans to work at depth is *Studies of Divers' Performance During the Sealab II Project,* by Hugh M. Bowen, Birger Andersen, and David Promisel, dated March 1966. It is available online at http://www.dtic.mil/dtic/tr /fulltext/u2/630518.pdf.

John Clarke's novels include *Middle Waters* and *Triangle,* the first two books of what he promises will be a trilogy. In his author biography, he describes himself as "an adventure-loving scientist and aviator who challenges preconceived notions about mankind and the universe." His fiction is great fun, and the diving science presented in the pages is both accurate and interesting.

CHAPTER 5: SUBMERGED

John Lethbridge's words come from a letter he wrote that was published in the September 1749 issue of *Gentleman's Magazine*. The Lethbridge suit has been described in many sources, but one excellent description can be found at http://www.devonheritage.org/Places/Newton%20Abbot /JohnLethbridgeofWolboroughandhisdivingmachine.htm.

For more about the minerals found in deep sea vents, see the brochure *Polymetallic Sulphides* at https://www.isa.org.jm/files/documents/EN /Brochures/ENG8.pdf.

Of the approximately 26,000 active duty submariners in the U.S. Navy, about 4,000 are officers and 22,000 are enlisted personnel. The average age of the officers is thirty-three and of the enlisted personnel twenty-seven. The Navy has begun to integrate women into the crews, and despite the challenges of bringing women into the close quarters of submarines that were built decades ago, there are, at the time of this writing, 121 female officers and 170 female enlisted personnel serving in the Navy's submarine fleet. All of these numbers were provided to me in an email from the U.S. Navy in 2018.

The full title of Dan Gillcrist's book is *Power Shift: The Transition to Nuclear Power in the U.S. Submarine Force as Told by Those Who Did It* (2006, iUniverse, New York). The book is a collection of lightly edited interviews with submariners, including both officers and enlisted personnel. Don Walsh wrote the foreword and was among the many retired submariners interviewed for the book. The estimate of time spent underwater on deterrent missions (that is, by ballistic missile submarines) appears on page 215. This estimate is based on statistics provided by the Public Affairs Office of the Strategic Systems

Programs, along with the assumption of an average of sixty-five days per patrol. It is at best a very rough estimate, and is, of course, a low-end estimate, because it does not include fast attack submarine missions, the missions of other nations in nuclear-powered submarines, or the missions of the thousands of diesel-electric submarines that operated in the past and that continue to operate today.

Boyle's words about Drebbel come from his *New Experiments Physico-Mechanical*, published in 1660. The story of Drebbel's submarine (or submarines—he seems to have built several) has been told and retold, and may have grown in the telling. One of my favorite sources is an article titled "The Origin of the Submarine," published in *Blackwood's Magazine* in 1917 (vol. 202, pp. 106–17) and available today at http://todayinsci.com/D/Drebbel_Cornelis/Drebbel-OriginOfSubmarine.htm.

Although the design of the *Turtle* has been lost, illustrations supposedly representing the little submarine suggest its basic layout. In 1997, students at Old Saybrook High School in Connecticut built a replica. For a video of the replica's test dive, see https://www.youtube.com/watch?v=AmNs5UwxCWE.

There are many, many excellent sources on military submarines. Aside from the aforementioned *Power Shift*, two that I found to be especially useful are Brayton Harris's 1997 book, *The Navy Times Book of Submarines: A Political, Social, and Military History* (Berkley Books, New York), and the 1993 book *Submarine: A Guided Tour Inside a Nuclear Warship*, by Tom Clancy (written with John Gresham) (Berkley Books, New York).

The Argentine submarine *San Juan* was found while *In Oceans Deep* was being finalized, as noted in the epilogue.

James B. Sweeney's 1970 book, *A Pictorial History of Oceanographic Submersibles* (Crown, New York), provides an excellent overview of submersibles up until the date of its publication. It includes not only 420 drawings and figures but also substantial descriptions of submersibles through the ages.

CHAPTER 6: THE ROBOTS

An article called "The da Vinci Robot," written by Michael E. Moran and published in 2006 in the *Journal of Endourology* (vol. 20, no. 12, pp. 986–90), begins as follows: "One might assume from the title of this paper that the nuances of a complex mechanical robot will be discussed, and this would be correct. On the other hand, the date of the design and possible construction of this robot was 1495, a little more than five centuries ago." The article explains that the robot—not a word that Leonardo da Vinci would have used, as it was not coined until 1920, when it appeared in the play *R.U.R.* by Czech writer Karel Capek—was controlled in part by a programmable analog computer. The topic apparently interested endourologists (who today use robotics in surgery) as an example of early robotics.

The 1975 book *No Time on Our Side* (W. W. Norton, New York) was written by Roger Chapman, one of the two men trapped in the disabled submersible until rescued by CURV III off the coast of Ireland.

I viewed the propaganda film about CURV in April 2018 at http://cyber neticzoo.com/underwater-robotics/1965-curv-cable-controlled-underwater -recovery-vehicle-jack-l-sayer-jr-american/, but it was no longer available online in January 2019.

Christopher Swann's 2007 book, *The History of Oilfield Diving: An Industrial Adventure* (Oceanaut Press, Santa Barbara, CA), offers an absolutely amazing detailed history of oil field diving. The book, based largely on interviews, includes hundreds of photographs and quotes from those who pioneered oil field diving. If there is any single book that I would recommend for anyone interested in oil field diving, this is it.

I could not locate any record of the Bondi colloquium on diver safety and ROVs, but Alex Kemp's 2011 book, *The Official History of North Sea Oil and Gas*, vol. 2, *Moderating the State's Role* (Routledge, London), includes several paragraphs highlighting Bondi's interest in the topic, and Bondi's 1990 autobiography, *Science, Churchill and Me* (Pergamon Press, Oxford), includes the two sentences used in the text expressing his interest in the potential for ROVs to gradually replace divers.

The figures regarding the number of ROVs and related information come from *The ROV Manual: A User Guide for Remotely Operated Vehicles,* 2nd edition, by Robert D. Christ and Robert L. Wernli (2014, Butterworth-Heinemann, Waltham, MA). This excellent and reasonably priced book should be read and read again by anyone using ROVs or supervising their use. Others may find it heavy going, as it is a technical manual, but it is by far the best reference I have found on ROVs.

The shipwreck found by Kaikō was the passenger and cargo vessel *Tsushima Maru,* sent to the bottom by the USS *Bowfin* during World War II. Tragically, at the time she carried, among others, 834 schoolchildren, of whom 775 died. The crew of the *Bowfin* would not know that the *Tsushima Maru* was carrying children until twenty years later, well after the war had ended.

The adventures of Kaikō have been reasonably widely reported, but some (probably most) of the literature is in Japanese, a language as far from my grasp as, say, a free dive to two hundred feet. One source of interest, in English and available at http://nsgl.gso.uri.edu /hawau/hawauw92002/hawauw92002_part5.pdf, is the wonderful paper "Development of a 10,000 m Class Deep Sea Research ROV Kaiko System," by Shinichi Takagawa, Taro Aoki, Kazuo Watanabe, Akira Takaobushi, Yoshihei Abe, and Katsuyuki Suzuki of the Japan Marine Science and Technology Center, which operated Kaikō. The technical note "Revisiting the Challenger Deep Using the ROV *Kaiko,*" by James P. Barry of the Monterey Bay Aquarium Research Group and Jun Hashimoto of Nagasaki University, was published in the *Marine Technology Society Journal* in 2009 (vol. 43, no. 5) and is available at http://www.mbari.org/wp-content/uploads/2016 /01/Barry-and-Hashimoto-Challenger-deep-2009.pdf. This report includes photographs and reads more like an impassioned adventure story than a technical note. I did not uncover any accounts of the emotional impact felt aboard *Kairei* when the launcher surfaced without the vehicle.

The aforementioned *ROV Manual,* by Christ and Wernli, includes a section on tether drag. Drag coefficients are dimensionless values that measure the relative resistance of objects passing through fluids. At one stage in his career, Christ empirically tested various tethers, assessing what amounted to drag coefficients. Based on comments he made during our discussions, I believe that he would agree with my assessment of tethers limiting the usefulness and maneuverability of small (and, for that matter, large) ROVs. As noted in the text, drag can be overcome somewhat by skillful piloting, but it remains a limiting factor.

Bob Christ and his father, despite years of searching, did not find the submarine they were looking for. It eventually turned up during an oil industry survey far from where they, and everyone else, thought it should be.

CHAPTER 7: AN OCEAN IN NEED

Sargassum is the genus of brown algae that gives the Sargasso Sea its name. In the Sargasso Sea, at least two species of the *Sargassum* genus reproduce while drifting about on the surface. Other *Sargassum* species can be found attached to the bottom, but if they break loose, they will continue to live and even thrive as they drift about on the surface. The sargassum clogging Dean's Blue Hole on the day of my visit probably included several of the more than three hundred species. In recent years, what may be sargassum population explosions have led to beaches and bays being entirely clogged with masses of the floating plants. These explosions may be related to climate change and nutrient (sewage and fertilizer) runoff, according to some sources.

The most detailed accounts of the enslavement of Lucayo Indian and African divers that I came across appeared in the wonderful book *Deep, Deeper, Deepest: Man's Exploration of the Sea,* by underwater archaeologist Robert F. Marx (1998, Best Publishing, Flagstaff, AZ).

Don Walsh was quoted regarding the *Deepsea Challenger* dive in the March 25, 2012, *National Geographic News* article "James Cameron Completes

Record-Breaking Mariana Trench Dive," by Ker Than, available
at https://news.nationalgeographic.com/news/2012/03/120325-james
-cameron-mariana-trench-challenger-deepest-returns-science-sub/, and
in James Cameron's film about the dive. Some of Don's words may
sound like a deep sea public relations sound bite, but since they come
from him, I take them as entirely sincere.

Sylvia Earle, Don Walsh, James Cameron, and many others are Ocean
Elders. Their organization is described at http://www.oceanelders.org
/ocean-elders/.

Earle says that her organization, Mission Blue, has reached more than half
a billion people through social media. To learn more about her organization or to nominate a Hope Spot, see https://mission-blue.org/.

Don Walsh's letter expressing concerns about Arctic oil and gas development, which was sent to Secretary of State John Kerry, is available on
the Ocean Elders website, http://www.oceanelders.org/ocean-elders/.

Despite BP's experience with the 2010 oil spill in the Gulf of
Mexico, the company has reportedly made recent comments suggest-
ive of recalcitrant callousness. According to an April 2018 article
by Andy Rowell titled "BP: Oil Spill Clean-Up Operations a
'Welcome Boost to Local Economies,'" published online by Oil
Change International (http://priceofoil.org/2018/04/09/bp-oil-spill-clean
-up-operations-a-welcome-boost-to-local-economies/), BP argued in a
regulatory submission that "in most instances, the increased activity
associated with cleanup operations will be a welcome boost to local
economies." This statement, with the word "welcome" removed, may carry
elements of truth, especially if the externalized costs to the environment
are ignored, but if indeed the report is correct, it seems remarkable that the
company would choose to make such an argument while still functioning
under the shadow of the 2010 spill.

The Nautilus Minerals website provides an overview of the company's
strategy for deep sea mining: http://www.nautilusminerals.com/irm
/content/overview.aspx?RID=252&RedirectCount=1. Also see Brooke

Jarvis's outstanding article about deep sea mining, "The Deepest Dig," published in the November 2, 2014, issue of the *California Sunday Magazine* (https://story.californiasunday.com/deep-sea-drilling) and republished in the 2015 edition of *The Best American Science and Nature Writing* (Houghton Mifflin Harcourt, Boston). The article describes, among other things, one respected scientist's decision to work within the deep sea mining industry because of her belief that this kind of mining is inevitable and that insiders can make a bigger difference in terms of environmental stewardship than outsiders. I once felt the same way, and I have published a short editorial about my views on the matter and been interviewed about my opinions. To some degree, I still hold those beliefs, although after decades of working with industry, I have come to the conclusion that, for me, the personal sacrifices—the day-to-day defeats and ongoing frustrations—are no longer worth the occasional small gains. However, the many individuals and environmental organizations that work with various industries play an important role, complementing the role played by organizations that simply fight against industry activities. I admire all of those who work closely with industry, provided they work sincerely for the betterment of the environment and for the use of honest science. I can say from firsthand experience that this sort of work requires tremendous fortitude.

I paid, and by local standards grossly overpaid, the lobster fisherman in Haiti, and after our time together, I gave him a new pair of fins, a newish mask, and a new snorkel, all spares that I kept in my own diving kit. It occurred to me later that in doing this, I was encouraging overfishing. I was assisting him in the act of what Sylvia Earle called "eating [his] seed stock." I am not proud of that, but at the same time I do not know what else within my means I could have done to help him.

INDEX

Note: Italic page numbers refer to illustrations.

ABOUT THE AUTHOR

Bill Streever is the bestselling and award-winning author of *And Soon I Heard a Roaring Wind, Cold,* and *Heat.* He began his working life as a commercial diver. Later, as a biologist, he worked on issues ranging from climate change to the restoration of Arctic tundra to underwater noise to the evolution of cave crayfish. Today, with his wife and co-captain, he splits his time between Alaska and their cruising sailboat, currently in South America.